FRACKING

America's Alternative Energy Revolution

JOHN H. GRAVES, ChFC, CLU

SAFE HARBOR INTERNATIONAL PUBLISHING

FRACKING
America's Alternative Energy Revolution
by John Graves, CLU, ChFC

Copyright © 2012 by John Graves

Safe Harbor International Publishing
121 N. Fir St.
Suite C
Ventura, CA 93001

This publication is designed to provide accurate and authoritative information in regard to the subject matter covered. It is sold with the understanding that neither the author nor the publisher is engaged in rendering legal, accounting, securities trading or other professional services. If legal advice or other expert assistance is required, the services of a competent professional person should be sought.

ISBN: 978-0-9835731-0-4
First Edition

Cover and interior design
by GKS Creative, gkscreative.com

Printed in the United States of America

To Mary, who asked the question,
"Isn't fracking dangerous to the Earth?"

TABLE OF CONTENTS

PART II

JAKE

Jake runs a frac rig in the Permian Basin: 28 men working a 14,245-foot vertical hole in the Texas desert. He stands 6 feet 3 inches and weighs in at 240 pounds, with the wide shoulders and stance of a Norwegian fisherman. Jake looks right at home here on the rig, though you could just as easily picture him as a Marine, an NFL linebacker, or as a sixth man on an NBA team. A quietly confident man, he looks you straight in the eye as he speaks, his voice deep rather than loud to overcome the site noise. Jake has been on frac rigs for 10 years, since he left college. He prefers tools to books, yet he can carry on a "petchem" conversation with the firm's engineers.

I am with a small group of New York financial analysts that Jake is guiding around the working machinery of a frac rig. The portable derrick rises 80 feet, skirted with a working platform at 15 feet. This doghouse is where equipment and men work round the clock. The wire-service gang has been troubled with their downhole shots since 3 a.m. They've pulled their rig and "jewelry" out of the hole, replaced the charges, and as we watch, dropped it again. At 80 feet a minute, it still takes nearly three hours to get it into position at the bottom of the bore. "It" is a tube filled with explosive charges. Imagine a 45-foot-long gun barrel that fires hundreds of bullets in a tenth of a second—sideways. The charges explode horizontally out of the firing tube, through the hole casing, into the solid rock to a width of 18 to 24 inches.

This is one type of fracing—in this case, for oil. Once fired, the wiring rig is then withdrawn from the hole, and the frac crew goes to work. They first run acetic acid to etch the frac zone. Next, they flow water, with an admixture of sand and prop agent, or "proppant," down hole. The perfect mixture of these is crafted to force and hold open the fractured rock. As the liquid is pumped out, the proppant remains. A handful reveals finely textured green sand mixed with perfectly manufactured black Bucky balls. With these in place, the fracture area is held open. Once the water is extracted, Jake's crew can pull more oil from the hole. The hydrocarbons will be released from a diameter of as much as 300 cubic feet.

Jake explains their newest approach: staging. The geologists' telemetry tools, as sophisticated as anything at NASA, have described the oil concentrations deep below our feet, at 50- to 100-foot intervals. In a large air-conditioned RV-sized wagon to the right of the derrick, two crewmen monitor the entire depth of the hole. We can see live electronic images registering pressure, seepage, conductivity, casing shape, and a dozen other characteristics of the penetrated rock. Jake's crew will run 15 stages on this particular well. They will drop the wire, charge the hole, wash it, prop it, and pump it. Next, they fill a 50- to 100-foot length with cement, let it set, and start a new stage up hole. Once each level is complete, they seal it off and move up. They know how much oil they can expect from each stage, what effect each charge will have on the following stage, and where to place the next cap.

Surrounding us on the barren Texas flatland is an armada of huge trucks and their working gear. Sand trucks have offloaded their first load into the sorter—250,000 pounds. Next to it, coupled by eight-inch-thick tubing, sits the gel tank. Gel is one of the admixtures that help to keep the props in proper suspension. This hole will consume 3,500 gallons. Smaller sacks and tanks of dry and wet chemicals are ready to be added, totaling 400 pounds. Together, the gel and proppants are less than one percent of the "slick water." We use the gel to wash our hands; with its texture of dishwashing liquid or "udder butter," it is a perfect cleansing agent for the Texas topsoil that invades every crevasse of your clothes and body. The strong morning wind sees to it that the sand finds every nook and cranny, fills every pore; in the Middle East, it would be called a *meltemi*.

Jake explains to the city boys (and me) the equipment noise-makers. He has three "triples and a single." These engines generate the horsepower to drive the entire operation, which includes the derrick, electronics wagon, the mix, and the pumping equipment. Rated to 10,000 horsepower, he will run them to 8,000, just below his required power generation. Off to the side, a 1,500-gallon tank of diesel fuel is connected to the engines. These roaring beasts are actually quiet. The pumping has yet to begin, as everyone waits for the wire-servicing crew to run and fire. Still, Jake uses his diaphragm to expel his words over the din of idling machinery. We ask our questions, and watch in childlike amazement. Jake can carry us across every conversation: rig workings, flow, operations, mixtures, financials, safety.

His entire crew prides themselves on 367 days without an accident. Jake attributes it to ongoing training, holding one another accountable, and the site's cleanliness. Everyone and everything is neat and orderly, despite the howling wind and painful flying sand. No one rushes about. Every man watches his every step and with good reason. If you are reading this on an electronic device, you'll see a video clip of a rig in action, and of men connecting a length of pipe to the drill string. They do this all day long, rapidly, with very few errors, though you can see the potential for danger. Everything on site is gargantuan, from the men to the gear, from the electronics to the casing. The very air is filled with latent energy of men and equipment.

To understand how commendable and improbable it is to achieve a year-long span without injury, let's step back in time, to two weeks ago, and see what's expected of the drilling crew:

- Dig a hole more than two miles deep to a pinpoint location, and in a less than two weeks, case it with 45-foot pipe lengths.
- Do so nonstop, with the drill bit rotating constantly to the right as it descends.
- Drill through 47 million years of time and deposits, through the deposits of the ancient sea that once covered this desert with nearly a mile of seawater.
- Pull the entire length of pipe out of the hole and replace it with casing to hold the hole open.

At this point, Jake and his professional team are called in to take over the site. They are the service guys, prepping each hole for oil extraction. They move across the desert from one hole to the next, quickly, in a cavalcade of 25 to 30 massive trucks hauling big gear. They are rewarded for efficiency; if they complete a job in 20 hours, they still get paid for 24. Twenty-acre sites continue to pull more oil from the depths than was ever imagined possible. Recovery rates of four percent from a well have always been regarded as the norm; "workovers"—when an existing well is redrilled—today can add three percent to that figure.

This "recompletion" rig cost $800,000 to punch and prep. At $85 a barrel, it will reach cost recovery in less than three years. Jake and his men will service it every few years, when they are called out. Gear fails. Holes seize up. Down-hole pumps break, and broken bits succumb to gravity. The guys speak often of fishing, prompting one of the New York City fellows to ask facetiously where the fishing hole is, since we are surrounded by sand. "You're standing on it," Jake says. Fishing means retrieving broken gear from the deep holes with specially designed tools. "When things break," Jake explains, "it's less costly to retrieve than to replace." The city boy takes all this in and from this day forward conjure a very different image when he thinks of "fishing." Jake is unfailingly polite and never condescending. He begins or finishes each sentence with "yes sir" or "no sir." For that matter, everyone we meet in this hellhole of a city deep in this dusty, miserable desert does the same, as is the custom in Texas. The reflexive words and basic civility are somehow reassuring. After all, it's 107 degrees out here; extreme heat tends to erode good manners.

A horizontal rig sits in the distance. Jake explains why it seems so far away—horizontal rigs cover a much wider area. They also cost more, and their technology changes with each hole. Mud motors drive the bit, free of the drill string. Once they get to depth, they will run horizontally 7,000 feet in a spoke array, covering far more of a deposit than their vertical counterparts and plugging and perfing in multiple stages. Horizontal drilling technology is evolving every day. "With the new rotary steering tools, we can shape to the reservoir, directly accessing more of the pool," Jake says.

Directional drilling is a more precise term, as drillers today can move in virtually any direction down hole, even upwards. The hole would look more like a snake's path through wet grass than anything else, measuring some 30,000 feet. Offshore, all this may take place beneath three miles of water, with astonishing accuracy thanks to seismic tools and the crew's skills and experience. Jake calls the offshore boys the "champs" of the industry. "That's why we in the field were so upset with the Macondo rig event," he says of the disastrous 2010 BP explosion in the Gulf of Mexico. He lists a number of mistakes made by management and higher-ups, including messing with the specs for the blowout preventers. "Any one of us groundhogs would have done it right."

We prepare to leave the site. I was impressed—with the technology, the applications, the engines and engineering, with the heat, and with the cleanliness. What I had not expected was the skill, knowledge, and courtesy of the men who run these rigs. They are professionals. They remind me of our young men in uniform, so proud, yet so quiet and sensible. Great respect is due these drillers. Jake and his men know more about their job than most workers. They have to know. Their lives and their reputation are on the line every moment they are in the field. Sloppiness, half-assed efforts, and ignorance, willful or otherwise, are not tolerated.

As we've seen, Jake is both knowledgeable and humble, a man who leads with quiet capability. As we leave, we also see in Jake something of a poet, and a philosopher. "You never know enough," he says. "This desert, that deep, old rock, always has something to teach. Each hole is different; each hole is a bear to win over. Guys who have been here for 30 years will tell you: *Every hole can break you. Listen for the lesson from each hole. Assume you can be killed or injured in the next minute.*

"You look at these fellows, these roustabouts," Jake continues. "If you met one on the street, you might turn away."

Would I? In this context they are consummate professionals. But they can be gruff, loud, and unkempt. Many are heavy drinkers and smokers. Criminal pasts are not uncommon. Neverthe-less, "They would give you the shirt off their back," Jake says. "They stand by each other. The rig team is like family. When we go home, we might call one another, just to make sure it's okay out there, out there in the other world." He doesn't call it the real world.

When they are done, the oil flows up and out to eight-inch flex hoses across the desert floor to storage, transshipment, and final processing into gasoline, natural gas liquids (NGLs), and a host of other chemicals locked in the hydrocarbon complex known as petroleum. These refined products are now the lifeblood of our economy. It all comes from age-old, single-cell creatures. It is brought to us through a manmade system of pumps and arteries that most of us are no more knowledgeable or aware of than the vulnerable workings of our own cardiovascular system.

For better or for worse, we are united with these ancient blood-lines. We are reliant upon rigs, pipelines, delivery systems, and men like Jake to maintain life as we know it. We are oil and gas.

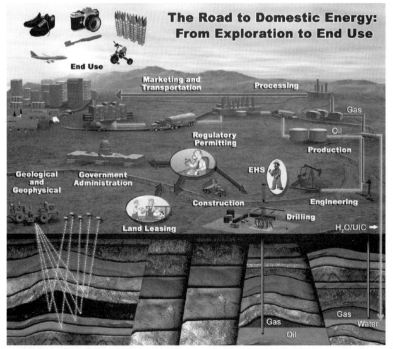

INTRODUCTION

In the energy business in West Texas, it's all about whom you know—from work, from play and from church. Family is important. If your father worked in the fields, you work in the fields. Your uncle graduated from University of Texas of the Permian Basin; you went to school at UTPB. Every man in the family, and many women, were members of one branch or another of the Armed Forces; so were you.

These links, this lineage, provides you with context—historical, psychological, and genealogical—and determines your prospects. Contacts beget other contacts. Wherever you work, you know someone, or someone knows of you. Whether you are a driller, a rigger or a developer, who you know defines who you are, and who you will become.

This tapestry allows each thread to stand out, while supporting every other thread. The ties that bind are strong. You sit in church a few pews back from the company president, and next to your fellow riggers. Your wives know one another. Their parents knew your parents, and your children go to school and play together. Your son may be the next company president.

The terrain of West Texas fosters this sense of community. This is one of the planet's deserts, one of the heat sinks of Mother Earth. The weather doesn't so much change you as it ingrains itself in you. Your body adapts to fit in. You may bring A/C and trucks and sustenance, but these also become part of the desert. You accept the desert as your life. Dry, hot, dusty, flat, scrubby. The Bedouin of another deep desert, the R'ub al'Khali, have an expression: The sand becomes part of your blood, never to leave. You become the desert.

As a visitor though, it makes you feel foreign. Flying in for the first time, you are impressed by the pockmarks dotting the white desert floor, each linked by an improbable road. It's like nothing you've flown over before—not the patchwork of farms in the Midwest or the tidy network of roads and rooftops delineating the suburbs. You will learn that these pockmarks are wellbore sites numbering in the thousands. Each one draws energy—oil, gas, and NGLs—from deep below the sand. Were you to fly over the Saudi desert, you would see a similar, vast array of energy source rigs. The view is similar over Taft, California, Calgary, Alberta, and a dozen other sites around the globe. Straws dropped down through the sand to the rich liquids below.

While we have burned petroleum over thousands of years and across dozens of cultures, the critical developments allowing access to the vast subterranean reservoirs in the shale fields are more recent. Technology, engineering, and human skills have brought forth access to these deep deposits.

The most recent of these has opened an entirely new realm of energy exploitation. The word "fracing"—pronounced "fracking" and often thusly spelled—has only recently entered our public vocabulary. It has been an oil and gas industry term since the late 1940s. It refers to hydraulic fracturing, or **the use of extreme pressure to force a highly salinized water and sand solution into an oil or gas well to open small fissures in the deep rock.** These fissures, held open by the suspended sand, allow hydrocarbons to flow through the pipeline to the surface.

Do you have any idea how oil flowing from deep below becomes gasoline for your car? Most citizens know it's a process that is politically charged and often controversial, yet few know how it works. For a fuller understanding, let's go back to our tightknit community in West Texas, start from the surface and work our way down.

The infrastructure of oil extraction is the jewelry of the Texas desert. Thousands of bright, small lights illuminate the night sky, flares pluming the evening air. Red-hot fire snakes up, evaporating into a thinning wisp of dark smoke. Rigs by the thousands take over small clearings in the desert. It is oppressively hot, windy, and dusty. Each well pad stands alone, its pumpjack—commonly called a donkey—pulling up oil or gas from deep below. Invisible lines beneath the desert floor connect the oil to reservoirs. These tanks hold the newest production for transshipment to the pipelines that lead to the refineries. These in their turn distill petroleum and gas into a dozen household, transportation and industrial products. Plastics, resin, paraffin, diesel, gasoline and aviation fuel are the proffered gifts from the deep. These rigs may also send natural gas to the pipelines for delivery to our electric utility plants. As more are converted to gas from coal and oil, the demand slowly increases for this simple ether. The pipelines also send natural gas liquids for processing into plastics and heating fuel and fertilizer.

This long thin line, this reed in the sand, connects our advanced civilization, indeed all of global society, to the planet's ancient past. These arteries bring forth the lifeblood of the global economy from rock deep below our feet. For better or for worse, we are indebted to oil, methane, and gas liquids. Try as we might, no substitution currently exists to replace hydrocarbon energy production. Nuclear, wind, solar, biofuels, and a host of alternative energy production options have been pursued, to little avail.

We should consider ourselves lucky on several counts:

Americans own their land and mineral rights, enabling an immediate sharing of the wealth of hydrocarbon energy production with farmers, ranchers, and landowners.
We have a century of developmental experience in drilling, extracting, transporting, storing, and converting this energy, which places us at the leading edge of further development.
The U.S. capital market flows freely still. Money seeks return, confined by the river banks of risk and regulation. The energy markets are ripe for development within these barriers.
Advantage flows to the planet as greenhouse gas emissions from burning natural gas are 45 percent lower than from burning coal. This gas recently replaced coal as the primary U.S. power source.

Let's explore the world of the driller, the fraccer, the trucker, the sandmen, the engineers and foremen, the field hands and managers, the capitalists and crew—all of whom together extract this rich liquid, this gaseous gold from the deepest voids. They find it, bring it forth, pipe it, deconstruct it, and ultimately deliver it in a thousand new forms for our use. However you feel about petroleum, you cannot deny your dependence on it. In the same way the desert becomes part of its denizens, petroleum is a part of each of us, and we are forever changed by it.

PROLOGUE

"America's Alternative Energy Revolution"—why such a provocative subtitle? Let's deconstruct these dangerous words.

The revolution will be televised. This is a revolution that is being broadcast with greater newsworthiness each day. Revolution equals change, a disruption of the old ways. As we shall see in Chapter 3, fracing is a disruptive technology. It is changing the way we source, extract, ship, store, and use the nation's energy. This revolution in energy sourcing is just beginning to become a revolution in energy cost. New power plants use less natural gas more efficiently at a lower unit cost. Our energy bills are beginning to reflect this. New plants are being built along the rivers of Pennsylvania and Ohio because of their proximity to the Marcellus gas source. New jobs are a result. The cost to produce fertilizer, aluminum, and finished metals is directly related to the cost of the energy used, and this cost is dropping. Imports of these materials—and of the crude oil that has been their energy source—will decline. Imports of crude oil from beyond the shores of North America are in steep decline, more than 40 percent since 2005. This is a revolutionary reversal of the export of our wealth overseas. In a decade, the US will be a major exporter of fuels, materials, and plastics. Refineries are being bought and refurbished along the Eastern Seaboard for this radical increase in supply of raw material: unconventional gas and oil. This revolution will take decades to unfurl. The youth of our nation today will be the revolutionary leaders of mid-century. This revolution may spread across the globe, much as democracy and capitalism have. The next generation of humanity may be born into greater wealth and opportunity. Who knows where the world will be by 2050?

An alternative energy revolution with hydrocarbons? Isn't that a misprint? Aren't alternative energy sources supposed to replace hydrocarbons? Aren't they the source of anthropogenic climate change?

Alternative energy sources are those that differ from the norm. The norm is crude oil, imported from abroad at great expense, used recklessly, without concern for the whole life cycle of cost accounting. The wise extraction of unconventional oil and gas from within our borders and its efficient use is the alternative to which this subtitle refers. We will learn of the dozens of firms and their technologies that attempt to more wisely exploit our own natural resource. Water use, reclamation, recycling, reduced use, or complete substitution are being pursued today. Methane capture at the wellhead is becoming a regulatory requirement. Reporting and listing of all chemicals used in fracing are spreading across the industry. Tax revenues—$65 billion in 2011—are increasingly devoted to road remediation and school, hospital, and municipal needs of impoverished counties recently enriched with frac income. Companies and regulators continue research and local investigation into water discharge, micro-quakes, and noise abatement. No firm willingly exploits the environment for its own gain and succeeds for long. The public weal discovers these inefficiencies quickly. Information is too readily available for any one source to maintain an exploitative advantage. And surprise: Very few firms have any desire to rape the land. The industry's HSE work stands for "health, safety, and the environment." Pride is taken in incident-free days, accident-free months. Profit flows to the firm whose legal department is not bogged down with plaintiff cases involving negligence. The environment is a profit center, in whole life cycle costing terms. The workers, the community, citizenry and the environment earn a significant return on investment.

The substitution of shale gas for coal in power generation has, and will continue to have, a large and positive impact on the environment. Already, the U.S. has reduced its greenhouse gas emissions, or GGE, by an astonishing 450 million tons since 2005, according to the International Energy Agency (IEA). No other nation has reduced its emissions by a single ton. Europe's emissions continue to grow, despite their full-throttle attempts to control them. Why are we so successful, when we don't even have a comprehensive national policy of GGE reduction? Simple. We are substituting natural gas for coal in the generation of electricity. A third of the nation's electricity production now comes from the extremely efficient burning of methane in dual-cycle turbines. These turbines are replacing older devices. The EPA has made it a virtual necessity. From this unlikely marriage of two opposites—the EPA and the frac industry—come three offspring: lower carbon emissions, reduced utility costs, and declining imports. These effects are just beginning to be seen in the economic statistics from the U.S. Energy Department's Energy Information Agency (EIA).

This alternative energy revolution is happening in America for a few simple reasons. Landowners retain all rights to minerals found beneath their soil. Few other nations allow these rights; they are State-owned instead. Individuals make better decisions than States, in almost all cases. Landowners are willing to assume the risk of development of land in exchange for the bounty derived. We are quite willing to put this out to bid to the firm best able to extract the deep resource. These firms have more than a century of extractive knowledge, skills, and capital. Since 1858, oil and gas companies have been drawing from the deep rock, shipping it through extensive pipeline networks, processing it efficiently in plants huge and small, and then storing it for ultimate delivery to myriad end users. This complex of industrial, capital, and personnel activity is unmatched in the world. Ninety percent of the planet's energy resources are owned by the State. From the wise Norwegians to the profligate Nigerians, few individuals, companies, or nations have the expertise, information, skills, or legal rights that we enjoy in these United States. We enjoin the developers to be wise in their complex approach to our national resource. If they misuse their capital, the markets quickly punish them. If they misuse their environmental stewardship, the courts and local opinion quickly destroy their unfair advantage and shutter their shops. Enron, ExxonMobil, and BP have each paid dearly for their egregious errors.

This is an American Alternative Energy Revolution for these simple reasons. Only here, only now, only us. America will lead the world once again.

The author is indebted to a wide variety of people and sources in his exploration of this timely subject. The discovery began as a simple investigation. Master limited partnerships (MLP) are a legal entity unique to the securities world, the world from which our story emerges. These MLPs are quite good at pouring off significant income for investors. Retirees love to receive distributions. These MLP distributions have several interesting characteristics: They are largely tax-free; the companies must pay out virtually all of their net income annually; they are designed to do just that. While tax reporting can be somewhat cumbersome, the annual income is consistent and typically grows at a steady pace.

The MLP would appear to be designed with the Baby Boomer in mind. As their use expanded in my Registered Investment Advisory practice, I dug deeper into the business of oil and gas. This led me to the new world of fracing. While the process is as old as the author, its exponential explosion in use had to wait for the drive and determination of one driller: George Mitchell. His success where others had failed opened the deep shale deposits. Texans had known for decades of the resource within these deep shales, but no one knew how to extract it. Mitchell's doggedness revealed the process. At the same time, his drillers were learning how to direct the drill bit horizontally. Their seismic tools, and those of other important exploration and production (E&P) companies, became sophisticated computers, able to see farther and deeper, while simultaneously showing 3-D images of the shale. These three fairly independent events came together in a cataclysm. Turn the drill bit horizontally to follow the now viable deep seams of hydrocarbon. Frac the seams to open them. Allow the pressure of tens of millions of years to force the gas, oil, and NGLs to the surface. Rather than dropping a straw into a pool, the typical image of an oil well, this is like a soaking hose in your garden, but in reverse.

Between 1998 and 2007, the world of energy exploration and production was revolutionized. Today, we continue to witness this New American Renaissance, a rebirth of an old energy source: oil and gas. The rebirth—with Mitchell as the midwife—has reversed the energy production decline in America. We are producing more oil and gas today than we have for decades, reversing a steep decline. This has the immediate effect of reducing our oil imports from "savage nations." It has the added benefit of tens of thousands of new jobs created during the worst recession in nearly a century. This revolution is offering new tax revenues to local state and federal authorities.

In addition, it has brought about changes that were wholly unanticipated just a few years prior:

- The U.S. is becoming a gas exporter, after having spent billions on liquefied natural gas (LNG) import facilities.
- Today we have the second largest natural gas reserves in the world, after expectations of precipitous declines in gas resources.
- The conversion from coal to natural gas for power generation is progressing as fast as production and delivery will allow.
- Affected local communities are discovering massive new wealth after decades of poverty and population decline.
- Environmental changes expected to take decades are occurring in years; these are changes for the better.

My research included visits to well sites, where I developed a profound respect for the workers. CEOs I spoke with passed along a wealth of knowledge ,insights and opinions based on personal experience, both positive and negative. Environmental attorneys I spoke with focused on the legal concerns of local communities and often indicted the oil and gas firms responsible for spills, injuries, and losses. Indictments aside, these folks tended to believe that responsible fracing would open up opportunities for our nation.

My most important discovery was the absence of information on the subject of fracing for the general reader. Technical, professional, and political books were in good supply, yet, there is little for the public that is reasonably objective—unbiased by either the industry or extreme environmentalists. Into this void steps a courageous fool, your author. I fully acknowledge my lack of petroleum engineering or of community organizing background. I do offer this journey as an honest appraisal of the frac world by a concerned citizen.

We shall examine a few points:

- Fracing is a tool to extract oil and gas. Along with horizontal drilling and new seismic tools, these technologies have coalesced into the discovery of an entirely new energy source.
- The natural gas now flowing from Pennsylvania, Ohio, Oklahoma, Louisiana, Colorado, and Texas is powering the newest electricity power generators, known as combined cycle turbines.
- The oil flowing from North Dakota, Oklahoma, Colorado, Texas, and Louisiana is replacing petroleum imports from the Middle East.
- This gas is now replacing coal as the primary source of power generation for the nation.
- The environment is a direct beneficiary of the clean-burning energy from natural gas.
- The U.S. is on track to exceed the Greenhouse Gas Emissions standards set by the 1998 Kyoto Treaty, without our having signed the document.
- The U.S. has reduced its "carbon footprint" by 450 million tons since 2005 as a direct result of fracing.
- While significant concerns are expressed about local environmental changes, the nation and the globe are in better health because of the fracing of natural gas and oil.

- State and local authorities work closely with industry to monitor and reduce local damage from the entire fracing process.
- Despite a recent public hue and cry, the Environmental Protection Agency (EPA) has validated many of the controls in place to ensure minimal environmental impact.
- Water use is a challenge being faced daily; its use is being reduced as new technologies bring new approaches to conservation efforts.
- The men who work the rigs in the Badlands, the Panhandle, the swamps, and the bayou are the "New American Heroes."
- These "big boys working their big toys" live in the most dangerous places on earth, doing the most difficult job imaginable, to the highest safety standards in the world.
- Accident and injury rates are in decline, and the industry has one of the highest safety standards in the world today. The results are impressive: annual reductions in reported injuries despite annual increases in rig counts, wells drilled and pipelines laid.
- Jobs in the industry go begging: Shale gas and oil development has created 1.75 million new jobs in the past five years. Another 600,000 will be needed by 2020 and an additional one million by 2035. Tax revenues during 2012 totaled $62 billion; by 2035, these will rise to $2.5 trillion. That same year, $5 trillion in capital will have been invested in shale and tight formations resources.
- Local impact fees in Pennsylvania, for example, have raised more than $204 million during 2012. These will help local communities to deal with the sudden explosion of welcome growth.[1]

- The U.S. dependency on foreign oil has declined precipitously: Less than 40 percent of our oil is now imported from outside of North America.
- We are net exporters of diesel and kerosene. We will soon be exporting liquefied natural gas (LNG).
- Our coal use is declining far more quickly than imagined just two years ago.
- Local communities reap the financial benefit of this new energy harvest: Landowners lease their mineral rights; communities garner long-term tax revenue; and new roads, schools and community centers are springing up in poor rural areas for the first time in decades.
- Poor farmers, ranchers, and simple landowners are receiving monthly distribution checks, making them newly wealthy, newly empowered, and newly enfranchised in their community's future.
- Oil and gas industry companies are working together with environmental groups to develop real solutions to the problems inherent in such an evolving industry.
- The consensus approach to whole life cycle systems management is both enriching and empowering to all involved.

These points address the macroeconomic world of energy source and use. While they cannot be denied, they can be disputed.

This book's intention is to help the reader get beyond dispute to resolution. Conversations are encouraged between the communities affected, as well as industry, regulators, and all interested parties. "We can…" is always a starter for any discussion. "Stop…" causes rancor and dispute. For the sake of progress, let's begin with "We can…" and go forward.

PROLOGUE

The global energy field is changing rapidly, and in ways no one would have thought possible in 1990, or 2000. In 2008, the south Texas shale gas fields had yet to be exploited; similarly the Niobrara fields in Colorado. The North Dakota and Pennsylvania fields were just beginning to take shape. The Arkansas and Louisiana twin fields were just coming to life.

In less than five years, the entire global structure of energy production and distribution has been altered. Globally and at home, the shale oil and gas "plays," or areas where gas or oil fields are in development, have become the major drivers of energy policy.

These plays are the new game in town. They are upsetting the carts for virtually everyone: from the Qataris, who were betting serious capital on increased LNG demand in America, to the oil majors, who have taken deep capital positions in offshore exploration and production (E&P), to the policymakers of many major nations who have driven the alternative energy capital structure. Each of these capital positions now appears to be resting on the shifting sands of the shale oil and gas revolution. Will they slide off, be consumed by the deflationary price implosion potential, or survive in altered states? This Renaissance at home is a revolution abroad. Witness these current affairs as they are happening.

Europe is belatedly bringing gas pricing into the 21st Century. The Qataris have turned on a dime and supplied their excessive LNG to the quake-damaged Japanese. Alternative energy is in intensive care while regulators in the U.S. and Europe fret. Gas-based electricity production, which was insignificant in 2009, is poised to surpass coal as the primary source for power production in the U.S. The U.S. will have exceeded its Kyoto "obligations" for GGE reduction by December, or earlier—another impossible dream. LNG will soon be exported from the U.S., while the Australians will take over from the Qataris as the largest LNG suppliers globally in the same time frame. Russia clings to its "take-or-pay" gas contracts whose pricing is tied, inexorably and impossibly, to that of oil. Israel has a massive gas and oil field just offshore.

These global policy issues are tearing the fabric of energy production. Assumptions about consumption are changing. Capital pricing models are being shredded. Contracts change before ink dries. Knock on effects at the national level will grow for every producer and user of energy. Can you imagine (like John Lennon) an OPEC meeting in it new Jerusalem headquarters in 2024?

Our biggest challenges here at home are connectivity and pricing—tying all the wellbores together and surviving the current pricing hurricane for gas. The real challenge in energy production lies not in the policymakers' boardrooms but in the oil and gas fields. Pipeline demand is for at least 250,000 miles of new pipe over the next two decades. Processing demands are driving the recent "cheap capital" acquisitions of Pennsylvania refineries by private capital, sponsored by the federal government. No one wanted these refineries. Now they are up for auction to the highest bidder.

Coal-fired plants that have recently been brought up to EPA standards for electricity production are at a significant cost disadvantage. Their cost to produce a kilowatt hour of electricity from coal is four cents; from methane it is two cents. A local Ohio mega power producer that has spent nearly $2 billion for pollution controls on a large plant will now shutter the facility. Gas is cheaper, cleaner, and easier on the equipment. Another coal burner in the area has reduced its coal purchases by nearly a third in four years—that means burning 20 million tons less coal today than they did in 2008. The invisible hand of the market allows capital to flow to its natural level. It seeks return commensurate with risk. Regulatory behavior meant to "clean the air" has cleaned the clock of these energy producers. They must turn to the market to replace this spent capital by reducing their energy production cost by 50 percent. The environment is a natural recipient of an unexpected cause. Thank you, Adam Smith.

Phil Auerswald's book, "The Coming Prosperity,"[2] reflects the renewal, yet again, of the spirit of America: entrepreneurship, property rights, and community-based decision-making.

This coming prosperity is the New American Renaissance. Entrepreneurs, landowners, and local communities are at the forefront of these changes. Capital decisions made in Houston and Pittsburgh have to be ratified in Lavaca and Homer City. The Renaissance spreads capital across Midwest America. It may do so for much of the world. The nation will certainly change.

This amazing story is unfolding even as you read this. Production of shale gas and oil and NGLs continues to increase—and the rate of increase shows no signs of slowing down.

Affordability of energy for every human born today may be, in 20 years, an inalienable right. The expected energy demands of 21st Century Homo sapiens may well be met by the excretions of the smallest creatures ever to have roamed the Earth. These phytoplankton and zooplankton of the distant eons have dropped in the hundreds of trillions to the seabed surrounding Pangaea. Covered and comfortable, they have been pressed by **lithification**—time and pressure—into the folds of the earth. Coral and shell have added their droppings. The seabed rises and falls, twists, turns and rolls over upon itself. Such stirrings taking place over 100 million years have turned these animals into methane, ethane, propane, butane, and oil—the petroleums and gases which are today being recovered from these ancient resting places[3]. Their natural progression to the surface, via porosity and permeability, is today greatly accelerated by petroleum engineers and field hands.

Since 1948, they have been coaxing these fuels from their deep rocks with the first fracing experiments. Since 1995, they have been doing so in massive quantities, thanks to George Mitchell's perseverance. Hydraulic fracturing and horizontal drilling open these deep seams, encouraging the gas to work its way through the pores between tiny rocks, urging permeability where none was thought possible.

The production and distribution of this energy source is changing every day.

- Enormous discoveries lead to massive extraction techniques.
- Global demands are forcing this stranded, regional commodity to be rebranded internationally.
- Importers are becoming exporters.
- Contracts are being rewritten as they are signed.
- Capital is being invested on scales unimaginable until recently.
- Projects unheard of previously are becoming commonplace.
- Men in the field with the experience and knowledge of today's American drillers are rising to the top of the professional employment pyramid.

If you want to make good money, go to college. If you want to make great money, and make a difference, go to the oil fields. Start as a roustabout, become a derrickman, an operations manager, or an engineer. Veteran looking for a job? Go to the fields of North Dakota, Ohio, Pennsylvania, Louisiana, Arkansas Oklahoma, Colorado, or Texas. There is plenty of work for those not shy about sweat and grease, willing to work long days for good pay.

This book is divided into two sections, with relevant chapters.

Part One offers an understanding of the recent story of energy resources, demands, and production in the United States.

Chapter 1 will explore our energy situation today. What are our resources, our uses, and our demands? How do we source and exploit energy for a 21st Century economy? What impact does this have on our lifestyle and the environment?

Chapter 2 will review the history of oil and gas exploration and production. We shall focus on the history of fracing, from its commercial origin in 1947. More than 2.5 million wells have been fraced in the U.S. since that time. What are the results?

Chapter 3 will explore global energy resources and the complexities that drive the global market. Is there such a thing as a global market with a shared interest, or are there fractured markets responding to differing needs?

Chapter 4 gives an overview of current and potential energy production arenas. What are the sources of energy production in the U.S. today? What are the potentials for future growth?

Chapter 5 takes us on a journey through the fraccing process itself in great detail .

Part Two offers an understanding of the potentials and challenges ahead.

Chapter 6 will travel down the pipeline, watch its construction and discover what happens to the gases and liquids as they flow towards storage and use

Chapter 7 will look at alternative energy sources other than unconventional gas and oil.

Chapter 8 will examine the subject from the point of view of the author's alter ego, his Dr. Jekyll. As a financial advisor, my "interest" in this world of oil and gas exploration and production first arose upon seeing the income possibilities for retired clients. Learn more here.

Chapter 9 discusses the markets for energy in America and across the globe.

Chapter 10 addresses the challenges, risks, environmental concerns, and politics associated with fracing. This is the 'meat of the matter'. Controversy meets confusion as myriad political, social, cultural and global stakeholders share their views. Hold on to your assets.

Chapter 11 discusses the continued development of national oil and gas fields through the viewfinder of consequences, sustainability and application.

Conclusions are yours to make. While the author's prejudices will inevitably crop up at times, his goal is not to persuade but to provide you with information on the background, current knowledge, and future potential of fracing and the risks and rewards. You can form your own opinion. He does hope, however, that you or any other stakeholder—workers and regulators, concerned citizens and capitalists—won't draw an either/or conclusion. To oversimplify a complex issue with Hegelian dichotomies—a choice between two opposites—is a rude departure from a national history of debate and compromise between all interested parties. The energy future of the nation, the potential carbon-free future of the planet, depends on sound decision making.

This book is the result of serious inquiry, first as a financial advisor and then as a curious, caring world citizen. The essential starting point for any inquiry is "conscious ignorance": We know what we do not know. We ask ourselves what is missing from our understanding, and then fill in those gaps by exploring facts and opinions. In doing so, we may not arrive at any hard-and-fast conclusions, but we do develop our own viewpoint. Distinguish between fact and opinion in what you read, what you think.

Most of us are guilty of assuming that our judgment is impervious to bias. It is not. Doctors, physicists, engineers, journalists (yes, even financial advisors) are all prejudiced. We assume our way is the right way, and that we know more than we actually do. We ignore evidence against our viewpoint, and accentuate facts in support of our conclusions. **Data mining**—selective use of information to support a position—is one the world's oldest intellectual pretentions.

Stephen Jay Gould unwittingly demonstrated the point with his expose, "The Mismeasure of Man," a critique of Samuel Morton's craniometry studies in 1839. Gould inadvertently proved his own thesis—that Morton was selectively using data to affirm racist assumptions—by also selectively choosing data to "prove" his point. Decades later, further investigation showed Morton's data set was quite strong; however, it simply proved nothing—certainly not his own conclusions. Gould also committed the fallacy of *ad hominem* attack: accusing Morton of personal fault—racism—for his "scientific" work. While Morton's reasons for investigation were racist (seeking to prove that African Americans had lower intelligence), his research work itself was done to the standards of the day. Because someone is a fool does not necessarily imply that they are evil, ignorant, or stupid. They may be wrong, but this should not necessarily be seen as a reflection of character unless their prejudice corrupts their work.

Prejudice functions to prevent us from learning more than we wish to know.

There is also validity in following "gut feelings," which can amount to wisdom—not always, but often. *Heuristic decisions* are those made on the fly, without recourse to logic or rational analysis. They are part of the "fight-or-flight" response encoded in our genes. American Airline pilot Chesley "Sully" Sullenberger proved the point with his lifesaving landing on the Hudson River a few years ago. He "knew" he couldn't pull off a water landing without power—but then, acting instinctively, did it anyway.

The facts presented here are important; so is your gut reaction to them. View this book as a departure dock, a loading zone for your intellectual growth and decision making.

Whatever conclusion you draw, you will no doubt agree that the world of fracing is fascinating. The people who do the work are among the brightest and best of our youth. Those who supply the capital that goes down the borehole have an almost an insane conviction. The people of the countryside whose soil is being turned over in search of deep wealth have received ample reward for their sizeable risk. The local and state regulators that monitor and develop the legal frame work within which these dangerous jobs occur, evolve personally and professionally each day. Those for whom the very concept of fracing is anathema should be heard. Their concerns are founded in the evidence of clear and present local danger.

The risks, while significant, appear localized and manageable, and the opportunities that fracing offers the nation are becoming clearer each week. Already, we have:

- Significantly reduced our petroleum imports
- Begun to change the power generation from coal to gas
- Made significant progress toward achieving the dream of the environmental community: a deep reduction in greenhouse gas emissions[4]

These three simple facts evolve directly from the fracing of gas and oil wells in the U.S. during this century. As these continue, our imports from the Middle East will dry up, perhaps by 2035.[5] Today, we hold the largest energy reserves on the planet, are the biggest producer, and natural gas is plentiful. If and when motor vehicles evolve to natural gas-based technologies, the ultimate goal of the most ardent Earth Citizen may be realized.

Industry professionals refer to the hydraulic fracturing of deep rock to release hydrocarbon as "shale gas development." The word "frack," though widely used by the media and general public, is poorly regarded by professionals involved in the process. If used, it should be frac. We shall do so here.

Now, let's get on with the story of shale gas development.

Beginning is easy. Continuing is hard.
—Japanese proverb

PART ONE:

Overview of Energy Resources

U.S. ENERGY RESOURCES

The energy resources of the nation are huge—Brobdingnagian

World Fossil Fuel Resources
America Leads in Recoverable Fossil Fuel Supplies

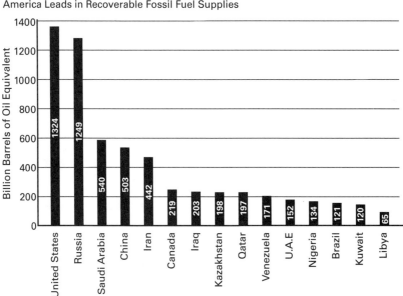

As shown in Figure 1, America's recoverable energy reserves are the largest of any nation. Three qualifiers are important: combined, proven, and recoverable.

- *Combined* means inclusive of oil, gas, coal, nuclear, and alternatives. It excludes such yet unavailable resources as methane hydrates and significant alternative energy opportunities.
- *Proven* reserves means "quantity of energy sources estimated with reasonable certainty, from the analysis of geologic and engineering data, to be recoverable from well-established or known reservoirs with the existing equipment and under the existing operating conditions."[6]
- *Recoverable* means "the amount of resources identified in a reserve that is technologically or economically feasible to extract."[7]

Within these limiting factors, we are the largest repository of energy resources in the world.

Shale gas, town gas, farm gas, wet gas, natural gas liquids (**NGLs**), tight gas, coal-bed methane—all are variations of the same basic chemical, methane, and its associates, ethane, propane, and butane.

- If it flows together with these, it is known as **wet gas** or NGLs.
- Tight gas is so named because of the difficulty in getting it out; its permeability is nil.
- Farm gas and town gas are so called because the locals get a portion of the production as part of their lease arrangement.
- Coal-bed methane comes from coal seams and is rarely used in the U.S. today, although it accounts for the majority of natural gas use in China and a few other nations.

Petroleum, gas, and NGLs may reside in a dome formation, lie in seams, or be fissured in highly compressed sedimentary rock. Each is the result of the deaths, by the trillions, of large and small plants, animals, and shell- or coral-based organisms over eons. The difficulty in extracting these hydrocarbons has led to the development of hydraulic fracturing technologies, horizontal drilling techniques, and advanced seismic search tools.

Figure 2 shows the exploitation of these resources:

Primary Energy Use by Source, 2011
Quadrillion Btu and Percent — Total U.S. = 97.5 Quadrillion Btu

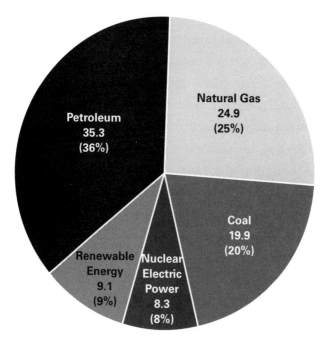

Source: U.S. Energy Information Administration, *Monthly Energy Review*, Table 1.3 (March 2012), preliminary data

Figure 2 is a snapshot of current use. The image has changed remarkably over the past five years. A quick review would show the decline in coal use, the increase in natural gas and alternatives use, and the virtual elimination of oil from electricity generation. By the time you read this, in 2013, the coal and natural gas figures will have changed places. The production of natural gas today exceeds all possible current national use. Storage units are filled as quickly as they are built. Exports of refined fuels already exceed previous records and LNG exports await federal approval.

The nation's shale oil and gas fields are only just beginning to be tapped. The deepest fields—Utica, for example, which extends from Ohio to New York and Canada— may take decades to fully develop. Technology evolves as quickly as it can be applied. You will read about sand, the simplest compound, and the nearly infinite variety of types, sizes, uses, and devices to which it may be exploited. Water use in the industry is perhaps the greatest ecological challenge. Its use is quickly adapting to recycling and reuse. With the newest techniques, water use may eventually disappear, particularly for the deepest, most water-intensive wells.

How do we utilize the energy we extract from the earth, the skies, and the waters of our country? Serious engineering skills, developed over several generations in both the laboratories and the field, have opened myriad forms of application. The simple water wheel of the 14th Century has evolved to the massive hydroelectric dams of today. These elegant, silent giants account for more than 70 percent of the nation's renewable energy use. During the past half century, the burning of oil for electricity has virtually vanished, replaced first with coal. Today, natural gas (yes, the methane of flatulence), is quickly replacing coal. As I write, the table has switched: In July 2012, gas exceeded coal as the primary source of electricity production in the nation, according to the EIA.[8]

Meanwhile, prices decline for natural gas, reducing the profitability of its extraction, and reducing the "rig count" for exploration and production (E&P) in many areas of the country. The "invisible hand"—a concept used by economist Adam Smith to describe a self-regulating marketplace— retains the power to direct economic activity, irrespective of think tanks, government spending, and the best intentions of many in the energy world.

Figure 3 shows what energy use looks like today:

Primary Energy Use by Sector, 2011
Quadrillion Btu

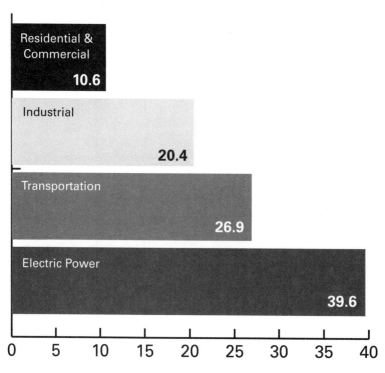

Source: U.S. Energy Information Administration, *Monthly Energy Review*, Table 2.1 (March 2012), preliminary 2011 data

As a nation, we use more than a quarter of our daily energy consumption for transport— cars and trucks, planes, and trains. Where does all this petroleum come from, this 27 percent of energy use that drives the transportation of the nation? Where do you think your most recent gas tank refill came from? If you guessed Saudi Arabia or the Middle East, you'd be wrong. Figure 4 shows the sources. Most of our petroleum is produced much closer to home.

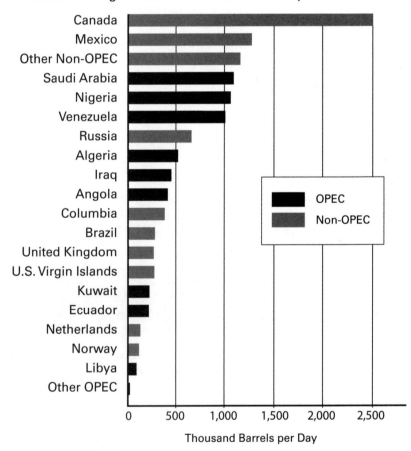

Sources of U.S. Petroleum Imports - 2011 Jan-June
6-Month Average = 11.5 Million Barrels Per Day

Thousand Barrels per Day

This chart is somewhat dated, referencing sources from 2011. Today, we import less than 40 percent of our petroleum from outside North America. Canada, Mexico, the Caribbean and, yes, the U.S. account for more than 60 percent of our petroleum use. The figure for imports from outside North America is declining rapidly and is expected to reach zero by 2035, if not earlier. This is almost entirely due to the hydraulic fracturing of shale oil deposits in the U.S. and Canada.

Technically, recoverable U.S. natural gas deposits are listed as 1,764 trillion cubic feet, of which 211 Tcf are proven reserves. Sixty percent of the 1.744 Tcf natural gas deposits are predominantly shale gas; tight sands and coalbed methane are a tiny portion. At U.S. production rates for 2007, the shale gas can supply the nation for 90 years[10]. The fields of Haynesville, Fayetteville, Marcellus and Woodford alone hold 550 Tcf of recoverable reserves, which at local production volumes is sustainable for several decades.

This sea of gas that we rest upon is in its infancy as a developing resource. Hydraulic fracturing was developed over the past two decades. Its cohort—horizontal drilling—has a similar history. The combination of these two technologies along with significant advances in seismic technology, and their application by men in the field, is defining the energy revolution of the country. One oil well pad today can draw upon the deep energy resources that 16 to 32 wells pads did a few years ago. That is an enormous cost savings. The reduction in local environmental impact is geometric.

Social Impact

What might this sea of gas do for the society that goes about its business above it? Imagine if our entire fleet of buses and trucks ran on natural gas. Today, many city buses do. A few long-haul firms such as UPS are experimenting with natural gas engines to replace diesel. The cost for a CNG engine is twice that of a diesel unit.[11] However, the fuel cost is at least 40 percent lower for natural gas, according to the EIA.[12] July 2012 prices show natural gas for trucks at $1.70 per gallon vs. $3.91 per for diesel.[13]

Look at the environmental impact we are currently seeing from the substitution of gas for coal in electric power generation:

- The U.S. consumes 22 percent of the daily global oil production—18 million barrels per day (BOPD).
- The U.S. GDP is 24 percent of the global GDP.
- We are the No. 2 source of greenhouse gas emissions (GGE).
- Yet, since 2005, the U.S. has reduced its GGE by a massive 450 million tons.[14]
- No other nation or group of nations has reduced their GGE. Everywhere these emissions are increasing, while in the U.S., they are in decline.
- By the end of 2012, we may have met the Kyoto Treaty requirements for U.S. GGE emissions reduction.

Let's look at the history of commercial natural resource operations beginning in 1947. The development, exploitation, and delivery of natural resources actually dates back to Edwin Drake's 1859 commercial oil strike at 69 feet below Titusville, Pennsylvania[15], but we'll only go as far back as Stanolind Oil's first experimental well in Hugoton, Texas in the spring of 1947. It used nitro to frac a well. The firm began commercial exploitation two years later.

According to the "Journal of Petroleum Technology":

Close to 2.5 million fracture treatments have been performed worldwide. Some believe that approximately 60% of all wells drilled today are fractured. Fracture stimulation not only increases the production rate, but it is credited with adding to reserves—9 billion bbl. of oil, and more than 700 Tcf of gas added since 1949 to U.S. reserves alone—which otherwise would have been uneconomical to develop.[16]

Shale Gas and Oil

In the past few years, shale gas and oil have become the key players in energy production in the U.S., with their production leaping between 2007 and 2012. Figure 5, a map of the U.S. from the EIA, shows each of the major gas fields: Barnett, Eagle Ford, Marcellus, and Bakken. The mid-sized fields are shown as well: Niobrara, Woodford, Fayetteville, and Hermosa Antrim.

Lower 48 states shale plays

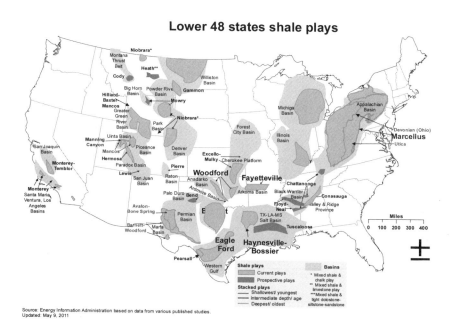

Source: Energy Information Administration based on data from various published studies.
Updated: May 9, 2011

Seventy percent of the nation either sits directly above these fields or has direct access to their hydrocarbons. Natural gas heats 56 million homes in the U.S. Burning natural gas supplies

32 percent of our power generation overall and 30 percent of our industrial power.

All of this puts people to work.

According to the Bureau of Labor Statistics, employment in oil and gas extraction and support services has increased by 27 percent since January 2010, and similarly by 11 percent since October 2008, when employment in the industry peaked prior to the recession.[17] Earnings have risen 15 percent, and at the same time, productivity has risen while workplace injuries have declined.

As shown in Figure 6, "shale gas plays," an industry term referring to geographic areas that have been targeted for exploration based on promising preliminary data, are abundant in the U.S.[18], according to the Potential Gas Committee, an industry-wide volunteer organization. Though they are from the industry, these volunteers generally don't stand to gain from their findings, so their work is presumably uncorrupted by self-interest or conflicts of interest.

Resource Category/Depth	Probable Resource	Possible Resource	Speculative Resource	Total Potentional Resources (1)
Traditional Resources				
Lower 48 States:				
Offshore (all drilling depths)	483,120	593,795	344,370	1,425,560
Offshore (all water depths)	16,820	52,060	61,780	130,000
Total Lower 48 States(2)	499,700	645,890	405,810	1,551,180
Alaska:				
Onshore (all drilling depths)	31,720	22,300	40,420	94,430
Offshore (all water depths)	5,140	19,500	74,790	99,370
Total Alaska(2)	36,860	41,820	115,130	193,830
Total U.S. Traditional Resources(3)	536,575	687,675	518,300	1,739,175
Coalbed Gas Resources	13,390	48,065	96,180	158,600
Grand Total United States(4)	549,963	735,740	614,480	1,897,775

1. All Total Potential Resource values, except Grand Total, are derived by separate aggregation and *not* by arithmetic summation of Probable, Possible and Speculative values.
2. Probable, Possible and Speculative resource values are derived by separate statistical aggregation and *not* by arithmetic summation of respective onshore and offshore totals.
3. Probable, Possible and Speculative resource values are derived by separate statistical aggregation and *not* by arithmetic summation of respective Lower 48 States' and Alaska totals.
4. Grand Total Probable, Possible, Speculative and Total resource values are derived by arithmetic summation of respective Total U.S. Traditional and Coalbed Gas values.

U.S. shale gas and oil production has increased 12-fold in the past decade. We are beginning to link many of the wells to their end users. U.S. utility firms use, or are quickly converting to, the burning of natural gas in the new combined cycle turbines. Petrochemical plants are converting to NGLs from petroleum for agricultural and industrial applications.

The impact on the economy of energy is becoming clear. Consumer prices for retail energy are dropping. Oil prices have stabilized significantly since 2007, while oil imports have declined. These declines can be attributed to four causes: the Great Recession; federal Corporate Average Fuel Economy (CAFE) standards for cars; car replenishment; and the abundance of shale gas in the U.S.

The first is, hopefully, a short duration effect. Demand typically increases as an economy recovers. This increase is less than robust at the industrial, manufacturing, or consumer levels. Companies have massive amounts of cash on hand as they wait to gauge post-recession demand for products and services. Nationally, corporations have sidelined nearly $1.8 trillion in cash reserves. Lending is stalled despite historically low capital borrowing costs as a direct result of this massive cash on balance sheets. Firms continue to strengthen financial balance sheets by paying off debt or replacing higher-cost debt with new, lower-premium paper—often of higher quality. Consumers remain tight with their money, either paying down short-term debt or increasing savings. The consumers' unsecured debt pile has shrunk dramatically since 2008, by more than 20 percent, according to the Federal Reserve.[19] Consumer credit card debt has fallen each year since 2008, in line with corporate debt reduction. Spending has barely kept pace with the "recovery" under way. We are far more cautious about dining out, or making that major purchase of a boat, new car or even a vacation. We have increased our contributions to our 401-k and 403-b retirement plans at work, paid down debt, and increased savings. These actions do not contribute to an increase in demand for energy generation or use.

The CAFE standards for vehicles are undergoing a rapid transformation. At the same time, the national auto fleet is constantly replacing itself despite our frugality of late. Natural attrition, vehicle upgrades, and the one-time federal "cash for clunkers" program have each contributed to newer cars in many garages. These cars are more fuel-efficient, weigh less, and are designed for better mileage. In the summer of 2012, at the time of this writing, gasoline prices have declined by more than 12 percent, which rarely happens during the summer driving season. Quite simply, the demand for auto, truck, and airplane fuel has dropped. Unsure of the future, we are spending less and driving less. Moreover, in the next 16 years, 74 million Americans are leaving the workforce, and retirees tend to use less gasoline and electricity.

The EPA has essentially outlawed the use of coal for power generation. However you feel about this political decision, the effect has been hugely positive for natural gas and the fracing industry.

Say what you will about alternative fuels, they have tripled their power generation—to 4.6 percent of national demand. Wind and solar, which are intermittent sources at best, depend on coal- and gas-fired utility plants for backup generation. Florida Power and Light (FPL) recently unveiled their star attraction: a 75-megawatt combined cycle plant. It is a solar facility tied to a natural gas plant, which supplies the energy when solar power is unavailable. FPL prides itself on its clean energy function, with 63 percent of the generation from gas, 20 percent from nuclear and a scant .06% from solar. Oil has disappeared from the radar as a source of electricity, at about one percent nationwide. Coal use is declining more rapidly than most industry insiders and observers would have thought possible, down to 32 percent in 2012 and plummeting.

In the energy world, natural gas has become the 800-pound gorilla in the room growing larger day by day. Natural gas is finally getting the attention it deserves, with both 2012 presidential candidates lauding it as an asset to our nation—a viable and valuable source of power, jobs, local wealth, and tax revenues.

The bulk of this is shale gas, the "unconventional gas resource." Its application expands constantly. Shale gas extraction raises concerns about water use, road erosion, micro quakes, and methane leakage; however, it is moving the country toward energy independence, low inflation goals, and greenhouse gas emissions goals. Not bad for such a young technology.

OUR NATIONAL ENERGY RESOURCES HISTORY

Do not go gentle into that good night...

—Dylan Thomas

As we have seen, the U.S. has the largest energy resources in the world. Unfortunately, there are no universally accepted definitions regarding our natural gas and oil reserves. "Hydrocarbon-in-place" describes the nation's totality of gas and oil, or the amount available in all the reservoirs in the country.

Proven reserves, or P1 for short, are those with at least a 90 percent chance of recovery with existing technology.

Probable reserves, or P2, are those whose extraction depends at least partially on price fluctuations in the world markets. In other words, if the price of the energy product increases, so does the probability of P2 reserves. Why? Price increases offer more potential reward for the greater risk associated with probable reserves as opposed to proven ones. The cost to extract further depends on field source, type of fuel source, depth, technology, contractual requirements, taxation, and regulatory impact. P2 reserves typically have a 50 percent to 89 percent probability of recovery.

P3 are possible reserves, with a range of probability of 10 percent to 49 percent. Energy product prices would have to drastically increase to warrant the further testing, exploration, and development of these more distant reserves.

The **recovery factor** is the amount of proven reserves that can be drawn from the ground. This can be estimated in one of three methods: volumetric, materials balance, and production-decline curve. Well production is assumed to decline over time, and the latter tries to account for both the reservoir capacity and the draw down. Often unexplained is how reservoirs can have more reserves today after decades of draw down. Please note the second appendix for a discussion of an entirely different understanding of the source of our oil and gas! Thus, each of these methods is "subject to annual review."

Depending on the field, the recovery factor can vary between four percent and ten percent. Price is the major determinant in recovery factors. The higher the price for the hydrocarbon, the more can be spent extracting it. Hard-to-extract becomes easier—capital "juices" the flow. As well, any given well recovery can be higher or lower, depending on the recovery technique, advances in technology, and equipment use. Fracing has had a major impact at this level. Drillers can go back to older wells, frac them, and draw more from the reservoir.

BOE, or barrels of oil equivalent, is a conversion unit used to compare gas to oil: 58 CCF (5,800 cubic feet of natural gas) equal a barrel of oil. **Btus** are British thermal units. **Tcf** means trillion cubic feet of gas. **EUR** is estimated ultimately recoverable hydrocarbons. It is historical production added to reserves figures for a well, region, or play.

Dry gas is methane gas in a greater than 85 percent concentration. **Wet gas** contains methane, along with greater than 15 percent concentrations of ethane, propane, butane, and trace minerals.

Having digested these important but difficult concepts, let's look at the recent history of drilling.

George Mitchell, the Father of Fracing

George Mitchell, 93, is the son of an immigrant Greek goat herder. His firm has been a leader among the independent oil and gas producers since the 1950s. Working his way through Texas A&M's renowned petroleum engineering school, he graduated in 1940 with degrees in chemical engineering and geology. After serving during World War II, he started his eponymous wildcatting firm, Mitchell Energy, over a drugstore in Houston, with working capital from Woolworth's heiress Barbara Hutton, local merchants, and a Houston gambler. In 1952, he hit one of the largest gas strikes ever found in the Boonsville fields in Texas. He licensed to supply natural gas to the city of Chicago in 1954, while beginning to buy up land in the Barnett. The firm expanded into real estate, offshore oil E&P, gas and NGL processing and pipelines.[20]

In the 1980s, George Mitchell began another of his visionary quests, this time for the shale gas beneath the Barnett. People laughed at his ventures into massive real estate development, gas processing, and others things. Each in its time hit massive pay dirt. Persistence was his strong suit.

Mitchell spent millions on his adventure. He took full advantage of the 1980 tax credit for "unconventional gas" exploration. He pushed his men hard. "He had a way of getting things out of people that they didn't know they could deliver on," said Dan Steward, his geologist.[21] He supported his field men, over the objections of Mitchell Energy's board. Dan Steward and engineer Mark Whitley knew of an accident in the field at a French oilrig. By mistake, too much water had been pumped down a wellbore, yet the well didn't blow. They knew that hydraulic fracturing worked with gel. They replaced it with a few chemicals and biocides. They put more pressure on the water than they thought it could stand. By 1998, their well was producing—to the tune of hundreds of thousands of cubic feet of gas. They had developed "slick water fracing," which they called light-sand fracing. Trial and error across two decades resulted in triumph, perhaps the greatest private investment ever made.

From the field, Mitchell refined the process, expanded it, and changed its technology. By the end of the decade, Mitchell Energy had figured out how to break open the shale and how to release the gas, oil, and NGLs. Slick water fracing became the new standard. The use of sand and water, normally a curse to any driller, was the answer. His firm also simultaneously developed horizontal drilling. This is the turning of pipe to the horizontal in order to drill through the center of the hydrocarbon concentration. Mitchell and others co-developed advanced seismic analysis with the advent of supercomputers. Now the geologist could see deeper, the driller could drill further, and the toolpusher could feel his way through the deposit to its core and set up the wellbore for the frac. They could do so in 2-D and 3-D imagery, in real time. They could monitor every inch of the wellbore for a dozen different variables.

George Mitchell detests the nomen 'Father of Fracing'. He sold his firm to Devon Energy in 2002, not for the money so much as for the capital intensity in further technology development. Horizontal drilling was Devon's strength. Mitchell saw that the combination of hydraulic fracturing and horizontal drilling would open the deep fields to an exponential increase in productive possibilities.

Devon Energy and Mitchell were the only firms in Texas that thought the combination was worthwhile. By 2003, they had 55 wells horizontally drilled and fraced. The technique spread like wildfire across the Texas prairie, into the Louisiana marshes and the Arkansas uplands. It then jumped state lines to New York, Pennsylvania, and North Dakota.

The sheer determination of a firm-handed man has created the single greatest boom in energy production in a century. The Greek goat herder would be very proud of his son.

Fracturing of underlying rock formations for the extraction of petroleum, natural gas, and NGLs has been contemplated and experimented with since the mid-19th Century. In 1947, Floyd Farris of Stanolind Oil and Gas Corporation (since acquired by Amoco) made the first effort at Hugoton in southwestern Kansas. Two years later, Halliburton Oil Well Cementing Company worked the first commercially viable wellbore. In total, 332 wells were "hydrofraced," with a production increase of 75 percent. By the mid-1950s, some three thousand wells were being fraced as news of the massive increase in productivity spread across the drilling world.[22]

The fluid today can be water, a gel (much like your dishwashing liquid), a propane gel, an acid wash, or a combination, known as slick water. It began as nitroglycerin, whose obvious danger forced a change to petroleum, as gelled crude, then refined crude. In 1932, water was tried, accompanied by the petroleum-gelling agents. By 1962, the first patent for a guar-and-borate gel was filed by Lloyd Kern. Guar is the gum in chewing gum. Surfactants and stabilizing agents are added; the primary chemicals besides borate are potassium chloride, alcohol, sodium chloride, and methanol. These have developed over the past four decades in response to well depths, drilling techniques, temperature and pressure, porosity and permeability of the local rock formation, and their functionality within the frac water. For example, etching the frac with acid (acetate) allows the proppant to remain in place. High-temperature wells require greater gel stabilizer such as methanol. Ultraclean gel agents and "encapsulated breaker systems" enhance the fracs' ability to remain open. Every well dictates a different approach.

Current Global Shale Gas Reserves and Their Exploitation

In 2011, the total value of the global shale gas market was estimated at $26.7 trillion.[23] Shale gas resource estimates will change over time as additional, more accurate information evolves. The international shale gas resource base is vast. The initial estimate of technically recoverable shale gas resources in the top 32 countries with shale gas resources is 5,760 trillion cubic feet (Tcf). Adding the U.S. estimate of the shale gas technically recoverable resources of 862 trillion cubic feet results in a total shale resource base estimate of 6,622 Tcf.

To put this recoverable shale gas resource estimate in some perspective, world-proven reserves of natural gas as of January 1, 2010, are about 6,609 Tcf, and technically recoverable world gas resources are roughly 16,000 Tcf, excluding shale gas. Adding the identified shale gas resources to other gas resources increases total world technically recoverable gas resources by over 40 percent to 22,600 Tcf.[24]

Technically recoverable shale gas in China has recently been estimated at 1,275 Tcf. Similarly large figures were recently released for Poland—3 Tcf of technically recoverable shale gas; however, once the results of a few wells were released, that figure has dropped precipitously, to 768 Bcf. These are only estimates. You can see how fluid these figures are, how variable are the estimates and their definitions.

While we are discussing China, their use of coal remains the primary source of energy production, given their vast reserves. Yet they have become a net importer, buying more than 200 million tons in for 2012. They also import nearly six million barrels of oil daily, a figure that may double in a generation. Their natural gas imports for 2012 were 58 percent higher than in 2011. Clearly, their demand for energy consumption grows, regardless of slowdowns or economic slumps reported by media. They produce no shale gas yet have a target of 6.5 Bcf in three years. This demand will be met.

Indonesia, Qatar, Australia, and the U.S. have LNG export facilities. These liquefy, transport, and re-gas shale gas for delivery. Among the most expensive capital projects in the world, their construction is attempting to keep pace with global demand. As the Chinese add their requirements to global demand, the LNG markets will expand significantly. An LNG ship may have a price tag of $200 million; an export facility may cost $10 billion. These are massive capital projects. Only the largest oil majors and major nations have the capital wherewithal to attempt such projects.

GLOBAL DEMAND FOR ENERGY

"Our world is headed toward what we at Shell describe as a 'zone of uncertainty,' a period of significant stress between energy supply and demand between now and 2050. Underlying global demand for energy is likely to double or even triple in the first half of this century. The world needs to invest heavily in energy production, both in traditional sources and in renewables."[25]

—Shell CEO Peter Voser

HS CERA Chairman Daniel Yergin has noted that the Western Hemisphere is emerging once again as an energy powerhouse. In his words: "Innovation is redrawing the map of world oil... and remaking our energy future."

These two men sit near the top of the oil and gas industry, one a senior executive, the other a trusted advisor to global leaders and well-known author on the subject of energy.

A tripling of demand for energy is a massive figure. The world currently consumes 85 million BOPD (barrels of oil per day)—the U.S. alone uses 24 million BOPD, or a quarter of this figure. The emerging nations striving towards the wealth and disorder of capitalism will drive this demand higher. The "next billion" people want what we have today. They will not take it from us; rather, they simply want to replicate it, without the attendant horrors of pollution, war, and family displacement. It is quite simple.

How they achieve, over the next 40 years, the lifestyle and livelihood that we enjoy today is the energy challenge for the human race. The resources required to produce 250 million BOPD of oil or their equivalents are available. The technology is at hand or in development. The knowledge, skill, and experience of workers across the broad spectrum of energy are in short supply. The appropriate regulation of energy evolves with political, social, and community demand. These events and people are real, honest, and challenging. All resources available must be utilized, as energy sources must be exploited, all technologies must be developed to meet this expansion in demand.

The population of the planet is leveling as it approaches seven billion. The aging of Europe, Japan, and the U.S. is occurring simultaneously with the leveling of Chinese demographics. In two generations, by 2050, the world will be far different. Its energy demands may have leveled off with its population. We cannot foresee either demographics or energy use; any forecasting would be absurd. We can continue to explore all energy resources. As our population levels, global energy needs will increase. The majority of the population will be urban by 2030. One billion of us are reasonably well off today. Another three billion are surviving. Two billion are pulling themselves out of abject poverty and starvation. The last billion remain destitute. These final three billion will migrate up the capital ladder—or die trying. As they do so, energy demands will grow inexorably. The next population explosion is of progression, of improvement, and of wealth creation.

This complex of interaction may be explored through the example of "Complexity Theory." A system is complex if it is **diverse,**

connected, interdependent, and adaptive. Complex systems are self-organizing. They emerge from the bottom up. The example is often given of a flock of birds or a school of fish. These do not have a hierarchy of command; there are no leaders, no apparent rules, and no boundaries.

Is energy production complex? Diverse it is, by definition. Interconnected, certainly by design. Interdependent, yes, it is engineered as such. Adaptive? It is so at each level we are considering. Complex worlds are self-regulatory, like a flock of birds or a school of fish. There are no leaders (other than those who misname themselves as such), yet there is discernible direction, willful movement, and intelligent design. Centralized control is as absurd here as it is in nearly every large-scale stratum. Its application simply breaks the units of cooperation down to chaos. The self-determinate nature of a complex system quickly reasserts its own "command-less'" function.

We can see this in global energy sourcing and its production. OPEC or the EPA may pronounce from their nostrums significant "remedies". The markets will have their day. The major corporations may try to sway direction but they are equally swayed by the group's direction. Dictators of capital and politics will eagerly open their mouths, ready to insert their foot. Their behavior barely ripples the water. Yeltsin, Khadafy, Khomeini, Chavez, and their "in crowd' of crusty cronies all make the media stage—their long-term impact on energy leaves small footsteps in the sand. Gasoline today is, after adjusting for inflation, the same price as it was in 1960 America.

We are left with a deeper respect for the invisible hand of Adam Smith's marketplace. His pin maker has become today's toolpusher. Demand, production, capital, labor, change, and innovation are the crowdsourcing phenomenon known as the global energy markets. Like a bird in the flock, a fish in the school, our advantage lies in participation and deep immersion.

The politics of international energy exploration and production is giving the U.S. industry at least a decade of advanced application. We have an open gas market that's called "gas-on-gas competition." It is priced daily and independent of the oil markets. The fundamentals of supply and demand dictate pricing in a very deep and liquid capital market. Virtually all shale gas here is available to any buyer, anywhere in the continent. The Henry Hub in Louisiana, where nine large pipelines meet, distributes gas to industrial and utility buyers everywhere. These two events—an open market and the radii of pipelines—define the pricing and delivery of gas in America. The market is open, and pricing reflects daily supply requirements. The gas is ubiquitous and anonymous; it comes from many sources, and is virtually identical from each.

European Energy Markets

Open markets and a confluence of pipelines are absent from all other sectors of the global market. Pipelines are spotty, at best, in Europe and Asia. European and Asian gas consumers have a "take-or-pay" contractual arrangement. Prices are indexed to the cost of oil, and minimum purchases are contractually agreed to for as much as 20 years in advance. The price fixing allows Russia to dictate the amount and price of its gas for much of Europe, from Ukraine to Great Britain. Valuing one raw commodity in terms of another is more than slightly odd, yet it is the system. Suppliers do not want to change. That would break the link, and free gas prices to float to demand tides. In Russia, 10 percent of their GDP - gross domestic product - is directly linked to gas contracts which supply 25 percent of the continent's gas needs. The largest energy producer in Germany, E.ON, pays a billion euros annually for gas that it does not need and cannot store.

Abrupt change would be anathema to producers, while saving utilities, industry and consumers. The European markets are having a difficult time adjusting to demand psychology. While change is rolling through the continent, with Henry Hub lookalikes appearing in Holland and Britain, they still account for less than 50 percent of the trade. Change comes slowly for many. The effect of a competitive spot-price market would certainly be more competitive bids and lower unit costs at both the spot and futures' pricing event. These changes are very difficult for the top-down managerial psyche found in both Strasburg and St. Petersburg.

Poland, knowing they are dependent on their neighborly Russians for energy, has approached the shale gas development very aggressively. They have experienced sudden drops in gas supplies during cold winter days, drops without explanation—unless one views this through a political lens. Many Poles do just this. The geological and legal challenges are significant. The gas lies more than 2½ miles below the surface. Legal rights to the land belong to the State. The original astronomical estimates for 5 Tcf of gas have been trimmed to size: 768 Bcf is still a fair figure. Majors such as ExxonMobil have recently exited the play, although their walk-away came after only two wells were drilled. The search continues in the continent's most politically open society, open to acceptance of unconventional gas resource exploitation.

The balance between supplier risk and consumer demand created by 20-year "take-or-pay" contracts has become a sclerotic neurosis—an inflexible obsession, at best. Consumers pay for policymakers' de rigueur dictates. The political web of influence, so well described by Daniel Yergin in his book, *The Quest*, has an undue, uneconomic impact on markets fraught with fraud potential. In addition to the Poles' experience of constriction of gas supplies, the Ukraine's attempt to resist a Russian gas cutoff on New Year's Day 2006 proves a point. The Ukrainian people suffered at the hands of their political masters in Kiev and Moscow for several freezing days before they agreed on a resolution. All of Europe saw the implicit threat. Cowering has become high art in Strasbourg.

The Qatari's massive capitalization of LNG of transport has begrudgingly edged the European markets toward gas-on-gas—a more open pricing market for gas. Their $17 billion gasification plants and their $200 million ships are delivering natural gas from the largest gas field in the world to all buyers. Gazprom, the Russian supplier, agreed to new contract terms in 2008. This was due certainly to the global financial crisis, yet after the above-mentioned cutting off of supplies to the Ukraine. Demand had dropped, and supplies had to follow suit. Today, as much as half of the European gas market is gas-on-gas.

Global Energy Markets

The global market remains quite multifaceted, pricewise. It may appear like a free-for-all today, with the most recent developments being quickly overshadowed by new LNG production facilities on the near horizon. Indonesia and Qatar are the leaders today. Their facilities are examples of major capital projects. It takes extraordinary planning, logistics, delivery, operations, and management acumen to design and construct these incredible works of capital art. The project requires the assured delivery of huge gas supplies, an entirely new plant to liquefy the gas, entirely new designs for very large, unique, ocean-going vessels, and another plant to accept and re-gasify the liquid delivery.

Recent demand within Asia has expanded rapidly, resulting from the entirely political decision to close down all Japanese (and German) nuclear power plants after the tsunami of 2010. The Qataris gladly supplied new gas, to the detriment of their slightly older European customers. Indonesia's LNG gasification plant, once the largest, is taking a backseat to others, while meeting local Asian demand. Australia is constructing a new $30 billion LNG facility. It will replace Qatar as the largest LNG exporter before the end of the decade. The Sabine Pass LNG plant in Louisiana is retooling to re-gas LNG, and export. A dozen other facilities are in the permitting stages for such export today. Indonesia, Algeria, and Malaysia are top gas producers locally. Gazprom wants to do the same with its Artic fields.

China expects to expand gas use from 6 Bcf to as much as 100 Bcf by 2020. Virtually all of their gas use today is coal gas, a by-product of their extensive coalfields. Much of their remaining use must be imported. Neighbors like Turkmenistan are more than willing to cooperate. Pipelines were installed between the two nations in record time.

- Global demand is one perspective. It can be estimated and projected with reasonable accuracy.
- Global production is another. Capital can be invested to bring fields, pipelines, and production facilities on line.
- Global delivery is a different animal. It remains a stranded market, a regional market, at best. Despite the efforts of Indonesia, Qatar, Australia, Algeria, Malaysia and Louisiana to make LNG processing a global delivery mechanism, it will take at least another decade for all of the pieces to fall into place.

A stable world economy is the prerequisite. Fair pricing of the gas commodity is also demanded from a deep and liquid market. Legal protections for all parties, across national borders, must be honored. Sufficient, capital-intensive delivery mechanisms must be built, and their costs must be reduced through use. Gas plants required on each end are expensive, as are the needed tankers. Capital projects of such magnitude demand political stability across such long time horizons.

A single LNG tanker has a price tag today of $200 million. The plants have billion dollar price tags. The price differential between markets must remain far enough apart to warrant this capital intrusion. Today, the figure for transport is $4 to $7 per MBtu. Asia prices gas at $18 to $20 per MBtu. In Europe, the price ranges from $16 to $20 per MBtu. U.S. prices today are about $3.75 per MBtu. Sabine Pass seems good to go, but for how long? Cheniere, the developer, is selling the delivery contracts to shippers, and is offloading the risk. Their price lock is a 15 percent surcharge.

Israel has the potential to become a world-class shale gas source, with an estimated 30 Tcf in shale gas fields offshore, and 500 million bbls of oil in the Valley of Elah. If these and other areas prove extractable, it vaults the tiny Middle Eastern nation into the No. 3 position globally for energy reserves, equal to or exceeding its erstwhile neighbor Saudi Arabia. With an education and scientific base equal to none, these resources will challenge OPEC and Russia as local sourcing of energy changes with the drill bit. The altered dynamics of the regional political equation will tilt inexorably away from an energy cartel with its origins in market-control dynamics.

Much of the global energy market is stated in dollars. In fact, the local nature of the fuel undermines the dollar figure. Gas and petroleum are actually priced at different price points depending on locale, supply, and availability in the global market. For example, "Brent crude" is one price and "WTI" is another, based on several factors: weather, supply chain, science, technology, politics and transportation availability. Beneath these two prices lies a far more complex layer of figures. North Dakota crude is slightly more costly than WTI and less costly than Louisiana crude. Why? Proximity of pipelines vs. train lines for transportation; directional flow of pipelines; refinery sites, functionality, and run rates. The detail gets broken down even further depending on the wells, time of year, and contracts for development and delivery. Entire doctoral dissertations have been written on the displaced pricing mechanism in the energy market.

For our discussion, know that the two generally accepted prices, Brent and WTI, are often erroneous quotes for a hodge-podge of daily pricing of crude oil. Turn this discussion to natural gas, and global issues arise. Russia prices its gas as a function of its crude price. This beggars their neighbors, the Europeans and the Ukrainians. The less-than-equitable market pricing creates a distortion in the energy universe. This distortion allows the U.S. to plan to export its gas, as LNG, to Europe and make a serious profit while simultaneously undercutting the Russian mobsters' price-rigging gamesmanship. Thus, supplier and end user are dancing around an ephemeral notion of "price." Price as stated in US dollars is an estimate, at best.

If political will cannot bend to market forces, it must snap. Prices will decline as supplies continue to ramp up both locally and globally. Stranding a nation by refusing to develop an available commodity may make sense to the Russians. It is economic suicide in the long run. Like it or not, globalization of shale gas is happening. Markets will respond, regardless of political decision-making. France, Bulgaria, and others have made their decision to ban fracing. They populace will bear the economic brunt of these decisions. There is no "one perfect" approach to satisfying the energy demands of a population. All solutions have some fairness and some irrationality built in. "Go forward gently into this dark and fearful night."

Alternative Petroleum Energy Resources

Alternative petroleum energy resources consist primarily of oil sands, or tight oil. While Venezuela has significant tar sands reserves, most reserves come from Calgary, Alberta, in Western Canada. With 1.7 trillion barrels of "bitumen oil" (think sticky, thick tar), and another 400 billion barrels of carbonate-trapped bitumen (currently impossible to extract), Canada has the largest tight oil reserves in the world. Recoverable reserves are 10 percent of this figure, putting them third in the world after Saudi Arabia and Venezuela for total oil reserves. Steam-assisted gravity drainage (SAGD) is the tool of choice today for extraction, at a cost of $55 to $65 a barrel. This oil will be shipped to America, through the controversial XL pipeline across our border, or to Asia via another pipeline to Vancouver.

Reserves exploitation is a direct result of pricing potential for a profitable delivery. Recall P2 and P3; the difference between these is primarily price. The state of the Chinese and U.S. economies today may give pause to further development of this Canadian resource. As the U.S. economy picks up, global demand will naturally increase. We still drive global production and demand for energy, as well as the more typical products and services. If the Chinese economy revives, it will compete with America for Canada's tight oil. Meanwhile, tight oil is expensive.

This respite should be short-lived. The breathing room will allow the political decisions to be made as to the direction of the pipeline for our neighbor's oil. Whether the pipeline stretches to the Pacific Ocean or the Gulf of Mexico, it will deliver global energy for several decades.

Extraction techniques are constantly subject to experimentation. The use of electromagnetic energy from large radio antennae sunk into the rock, dropping electric heating coils into the rock, or heating oil-based solvents and flushing them down into the reserves are each under review. If they work on a large scale, the use of water and the associated cost may drop by as much as 40 percent within the decade. This will bring the cost more closely in line with shale gas and oil—as long as the two fuel sources don't compete head-to-head. By that we mean, petroleum is used primarily for gasoline, diesel, and aviation fuel; gas is for power generation. If shale gas were to become a viable alternative to gasoline for automobiles, the oil sands play will be in trouble. Thus, it is an "alternative energy source."

The deep drillings of the South Atlantic and Western Indian oceans are yielding significant discoveries. The exploitation of these discoveries may lie a decade or more into the future. The massive capital outlays require the heavy lifting of the boundary nations of Brazil, Namibia, and Mozambique, or the deep pockets of the majors and their service company supporters. These "pre-salt deposits," while huge, are also the most difficult to get to and extract. Their development must await serious demand input to the global energy market. This will come. As we saw, this demand is expected to triple in 40 years. These resources may be the primary energy sources during the mid-century.

U.S. Shale Gas and Oil Resources

U.S. shale resources have developed rapidly over the past 10 years. Figure 7 shows the source of gas production from various U.S. shale plays during the first decade of the 21st Century. The increase in production is marked and varied. While the Barnett field continues to be the workhorse, eight other fields have added such capacity and production that they together have increased national production 12-fold. These figures are beginning to stabilize, as production responds to price depression. Hurricane Katrina in 2005 forced natural gas prices up to $15. Gas prices have since dropped to little over $3.50 today. Markets continue to function. Supply drives demand, which stimulates production and drives down prices.

Shale gas offsets declines in other U.S. supply to meet consumption growth and lower import needs

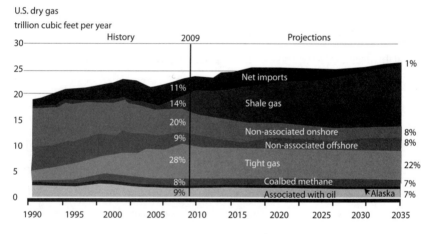

Oil Field Services

Oil Field Services

Oil field services (OFS) companies are the folks in the field who do the heavy lifting, like those led by Jake. They send down the drill string two, three, or more miles, then turn at right angles and run it out one to seven miles farther. Their accuracy is such that they could find a briefcase at 10 miles with this drill string. This is just the beginning. Schlumberger, Halliburton, Baker Hughes, Weatherford International, Transocean, and Cameron are each OFS companies. There are hundreds of small- and mid-sized firms in the U.S. and globally. These firms are the big dogs in the energy world today, but were the pups two decades ago. Their gross income for 2011 was $750 billion. One firm, Schlumberger, spends as much as the oil major ExxonMobil does on R&D—$1 billion annually.

Their independence has allowed state-owned enterprises (SOEs) to explore and develop without the aid of the majors, simply by outsourcing the work to the OFS folks. SOEs account for as much as 90 percent of the oil reserves in the world. Their decisions and actions dictate the global extraction and delivery of petroleum. While maintaining their close relationship with the corporate giants, the OFS firms have begun to joint venture some drilling risk with SOEs, often at the expense of the big boys. The oil giants still have the global management and systems expertise to design and manage huge projects, something neither SOEs nor OFS have. OFS continue to work the technologies of exploration. At $500,000 daily rates, offshore rigs that can save a few weeks in drilling time make serious money for their owners. This true in the gas fields as well. Get in, get down, get fraced, and get out. Engineering and information processing are the winning plays in the oil and gas game today. OFS companies plan to dominate the field and win the "big dog" competition every year for many years to come.

Economics of Shale Gas

A recent report from the Yale School of Management, "The Arithmetic of Shale Gas,"[26] demonstrates an attempt to compute the overall economic impact to the nation of shale gas production and use with the traditional Cost Benefit Analysis method. Using 2010 as an example, they illustrate a national economic impact in excess of $100 billion. In addition, the substitution of unconventional natural gas for 1 million BOPD of oil equals a national savings of more than $23 billion each year. This is the direct US balance of payments benefit of the new technologies of fracing, horizontal drilling, and new seismic tools. The report factors in the negative economic impact of spills, at an estimated cost of $2.5 million per spill, with 100 spills each year. According to this thesis, gas energy is far less costly to the U.S. economy and to the U.S. consumer than oil, or any other alternative energy source. In fact, the unconventional development of shale oil and gas is contributing nearly $1 billion dollars a day to U.S. GDP. This figure represents more than three percent of GDP in 2012 and will continue to grow.

The hydraulic fracturing of the deep rock is a 'disruptive technology'. This is one that radically changes a system into a new course, a new way to perform a task. It makes the universal suddenly becomes old fashioned. Remember your seventh grade math class? You were instructed in the use of the 'slide rule'. Do you recall how to use it? Do you know anyone who has one? It was the finest tool in the science class room. A giant one hung at the top of the blackboard and was taken down to use for precise calculations. It was an amazing tool. Its accuracy was strangely universal. Anyone anywhere could use it, irrespective of their language or culture. It was a primordial tool, like a hammer or a screwdriver. It was ubiquitous.

Ask you son or daughter about the slide rule. A look of discomfort, of gentle admonishment, will greet you. 'No dad, we use computers. That is a dinosaur'. The computer was the disruptive technology that erased the universality of the slide rule and made it a quaint curiosity. It made the slide rule the rock to today's finest hammer.

Hydraulic fracturing is a disruptive technology. It has, it is, radically altering the universe of energy extraction, production, shipment and delivery. It is the simplest of technologies when you look at it. Simply by turning the drill to the horizontal, you allow for the continuous extraction of an exponentially greater sum of hydrocarbon. Just as the substitution of binary calculations for mechanical ones disrupted the technology of local mathematics, turning the drill has disrupted the delivery of energy to the nation. Yet, it is far more than simply turning the drill bit. There are actually three disruptive technologies at work, simultaneously discovering, opening and delivering massive amounts of cleaner energy at far lower cost to an emerging economy, our economy.

These three technologies are **information technology, accessive technology and rendition technology.** The first is the application of massive computer processing to locally derived data. This data, as we shall see later, comes from a simple hammer. It is a very large and powerful hammer; yet, it remains just that. A 'thumper', or a series of thumpers, strikes the ground. the echoes are recorded. These become the data that speak of the wealth beneath our feet. In the past, a wildcatter might drill 100 wells and hit a gusher, as if by luck, with one. He had little idea where it was, what the volume of oil it held or how long it would last. Today, there are very few wildcatters left. Virtually every well spudded hits pay dirt. Geophysicists know how much is where and how long it will last. They know the characteristics of every foot of the hole before it is drilled. Once it is drilled, they know even more.

The application of processing power to loud thumping noises ultimately describes the depths of the earth in exquisite detail. This seismic technology tells the geologist exactly how much oil or gas or NGLs are at what depths and which are of primary and of secondary extractive importance. It shows the extent of the hydrocarbon layering, the shelves. Do this enough times, put the data into a 3D illustration and you have a map of the seas beneath your feet. Columbus had no such map of the seas before him, yet he discovered a New World. Imagine what we are discovering today. The stacking of deposits suggests repeated drawdowns of reservoirs – from the same pad and drillstring. Again and again and again. This is geometric growth. two times, three times, ten times re-entry and extraction from different stacks of hydrocarbon deposits. We literally do not know how much can be drawn from how deep. We can see what is there. Time and technology will determine our depletion rates.

The second disruptive technology is accessing the deposits, the fields. We will read about the Barnett, the Bakken, the Marcellus and others, during the course of this book. You will think of the geography of the USA in an entirely different way. Once these fields are discovered and described in such extraordinary detail, George Mitchell's simple, elegant solution comes to play. Turn the bit to the horizontal. The phrase in the field is turning the bit to the right. It turns to the right as it rotates down into the ground. Once you reach your depth, or just above it actually, you begin to turn the bit. We will learn about the process later. For now, simply imagine the vertical wellbore suddenly beginning the kick off to the horizontal. In about 1,000 of vertical movement the bit can be turned to the horizontal. This is the new accessing technology. A simple yet elegant solution. In the past, you would drill vertically to access the deposit. If it was extensive, you may have a hundred derricks within a square mile. A hundred straws drinking from the same fluid. Turn the bit to the right and you have one well drinking from all the deposit. Put a dozen wellbores on one drilling pad and you can access the entire deposit for miles in all directions from one very noisy, very large site. Sixteen wells can now drain from a deposit of hundred of square kilometers, from one pad. The applied technology of simply turning the drill bit to the horizontal is our second 'disruptive technology'. The economies of scale are enormous. They don't simply 'scale up'. They increase exponentially. this the same result as massive computing power in series: cloud technology for today's computers. The data flow capacity becomes almost infinite, as does the oil and gas flow.

Our third 'disruptive technology' is rendition. You have found the deposit and drilled down to access it. It has cost you millions to do so. You need to be smart about capital deployment. How can you pull as much as possible out of the deposit? Frac. Frac the deposit. Use explosive charges to open the neatly packed hydrocarbon. Examine a core sample from below. Hold it in your hands. It is solid, compressed, tightly woven sedimentary rock crushed by eons of time and millions of tons of rock above. Recall that these deposits are 5,000 to 15,000 feet down, or deeper. This is no longer sand on the beach. How do you open this densely packed rock to loosen the black residue you clearly see and feel? Frac it. You will learn how the process works in a later chapter. You fire explosive charges into the rock to open it a few inches, perhaps a foot or two. You are creating small fissures, nothing more. Then you pour millions of gallons of salty, sandy, water mixed with chewing gum, dishwashing liquid, cleaning detergent and some really nasty fungicides and bactericides (yes, they are lots of slimy little critters down there) under enormous pressure. The water forces open the rock for dozens of feet above, below and around the casing. These capillaries become the causeways of extraction. You pull out most of the water and leave most of the sand. The sand holds open these tiny freeways for thousands of miles of arterial length. Just wide enough for the smallest molecules of gas and oil to pass down—the pressure to escape is difficult to imagine. Once you have the entire length plugged and perfed, you drill through the plugs and let 'er rip. The IP, initial production, on these horizontal wells lasts for just a few months, but it flows more than a dozen vertical wells per second. You have rendered the hydrocarbon to the surface.

These three disruptive technologies have discovered, opened and produced the great gas and oil fields below. A century of extraction lies before us. You will read in the Epilogue your author's rendition of the near future. Science fiction? Probably. Yet, it describes the general domains of probability that lie ahead, if not the actual events. The ability to exploit the clean energy source within our national boundaries is due to the rule of law, capitalism and our apparently infinite resource of optimism. These gas and oil fields lie beneath many nations. Few have the requisite structure to extract this clean energy, few have the capital, few have the drive. If ever the table were set for a New American Century, this table we have before us is it.

U.S. ENERGY SOURCE PLAYS

Magic markets don't appear all the time,
so you take advantage of them when they do.
—Marc Andreessen, American entrepreneur

et's examine several of these primary energy sources in detail—local economic effects, reserves, expert analysis.

The current U.S. energy "plays" are listed in no particular order:

Barnett – Texas
Haynesville – Louisiana and Texas
Fayetteville – Arkansas
Marcellus – Pennsylvania and West Virginia
Bakken – North Dakota
Eagle Ford – Texas
Niobrara –Colorado, Kansas, and Wyoming
Utica – Ohio, Pennsylvania, New York, and Quebec
Woodford – Oklahoma

Lower 48 states shale plays

Source: Energy Information Administration based on data from various published studies.
Updated: May 9, 2011

These fields (see Figure 9) have come into play as both older oil fields that are in recompletion, and as new fields of hitherto unknown potential. The shales of the Ordovician and newer layers, or shelves, of sedimentary rock were never considered available until George Mitchell came along and figured out how to get below the rock, force and prop open fractures, and turn the drill to the horizontal. These three discoveries gave rise to a renaissance in American energy. His 30-year dream became a reality, first for his firm as prime mover, next for the industry as quick adaptors. Let's look at each field from the points of view of production, economics, employment, state and regional impact, and local importance.

There are a host of other fields small and large, known and unknown. These and others may take center stage as the decades unfold. New extraction techniques, more workers, and better safeguards will each impact future shale oil and gas development.

The nation is replete with gas. There are far more unexplored or underdeveloped fields than those in production. Entire areas of the Marcellus currently are 'out of bounds' for political and cultural reasons (New York and North Carolina, for example). These undeveloped energy sources for future unconventional gas and oil exploration and production will allow the flow of energy to continue to grow for decades. Examine for a moment a map of the gas pipelines crisscrossing America (see Figure 10). These pipelines come in at $1 million per mile today. There are more than 200,000 miles in place. You get a sense of the capital power engaged in American exceptionalism. You can see the dense layers of development in Texas, Oklahoma, Arkansas, and Louisiana. Also note the complete lack of pipelines throughout the Bakken and Utica, the extrusions moving toward the Marcellus and Niobrara. Far more wells are ready for transshipment than there are pipelines. The need is conservatively estimated at 250,000 miles over the next 20 years.

Natural Gas Pipeline Map

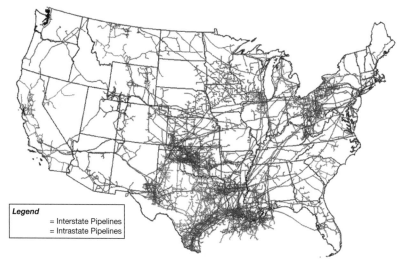

Source: Energy Information Administration, Office of Oil & Gas, Natural Gas Division, Gas Transportation Information System

Barnett

Texas Counties with Producing Wells

The Barnett (Figure 11) is the best explored and one of the largest shale gas deposits in the U.S. Its geology is well-known, but the overlay of shale has prevented exploration and extraction until recently. The current estimates are for more than 43 Tcf of gas. The Barnett provides for more than 119,000 jobs and $100 billion in annual economic activity, and $912 million in tax revenues.[28] [29] A portion of this activity comes from land leases of mineral rights to E&P firms. Landowners lease the right to extract the hydrocarbon and, in return, receive a portion of the royalties—typically eight to 12 percent. It and the Eagle Ford fields will soon be responsible for more than half of all U.S. gas production.

The Barnett is tight gas that must be extracted by horizontal drilling and fracing. Over 18 years, building on his pioneering experimentation, George Mitchell developed the light sand frac, using far less sand and less liquid under pressure. Today, we know it as "**slick water fracing,**"which we will explore in more detail later. Once the flows began, they were sizeable. Everyone in Texas knew the size of Barnett. Extrapolating from Mitchell's initial success, the industry saw by 2003 that this was going to be the next "gusher"—but of gas rather than oil. Mitchell sold to Devon Energy, which is the major E&P firm in the region, followed by Chesapeake. Gas and NGLs are being removed in record quantities, while rig count has declined with the price of gas. Dow Chemical has recently devoted $4 billion to a new NGL plant in Louisiana specifically for the NGLs from Barnett. The conversion to ethylene will support a thousand new jobs and create a value-added export play to the area. Fort Worth, Texas, sits on top of a portion of the Barnett, and is deeply involved in its development, along with the Texas Railroad Commission (TRC). TRC is the state regulator.

Water use is a concern in this arid part of Texas. The industry today uses an average of one percent of the local groundwater each year for fracing. This figure varies with each site. Rural sites consume more water than urban, and some sites consume more than others because of the wellbore characteristics. The TRC expects water use to grow to six percent of groundwater over the next decade. Recycling efforts are growing in response to these concerns, as well as to the cost of water purchase, transport, and use. Because much of the water recovered from a frac is brine (high salt content), its disposal is important to ranchers and urban populations. Injection wells force the brine back beneath the shale into the porous rock below with **disposal wells.** The brine is also recycled back into new fracs with **injection wells.** As Texas has the largest number of oil and gas wells in the nation at 216,000, it also has the greatest number of disposal and injection wells, at 50,000.[30] The drillers reuse what they pull back up out of the borehole or recycle it down these types of wells. The issue with disposal and injection wells is simple: The water is removed from the hydrologic system, permanently. This answer was sufficient when water use for drilling was less—before hydraulic fracturing. Today, these wells are viewed ominously, as threats to the local hydrology. Such a withdrawal has to be replaced with other methods.

As a result, an entire water-use industry is springing up as water needs increase. As the local communities challenge the increasing demand for water by the fracturing industry—particularly in drought-susceptible areas like the central Texas plains—new technologies are developed. In addition to the disposal and injection wells, biocides, electric processing, and "electrocoagulation" are in use.

Fountain Quail Water Management is an example of a firm that recycles on site. It has expanded its business in the Barnett. Their ROVER system is a mobile answer to water sourcing, recovery, treatment, transportation, and reuse. It can recycle 10,000 gallons daily, removing particulate and organic suspensions. The resulting brine can be mixed with proppants and reused.[31]

Firms such as **NeoHydro** also clean the brine at the wellhead. They treat the water without the use of potentially dangerous biocides, while removing heavy metals and solubles. Its "electro-oxidation process" has been successful with several wells. The firm claims that the cost savings over water shipment or bioremediation can exceed 70 percent. Clearly, the savings can be substantially more than the added cost of these new treatments.[32]

Halliburton's proprietary treatment process "CleanWave"™ uses an electrocoagulant to destabilize the brine through ionization. The suspended matter rises to the surface and is skimmed, leaving water for reuse in the well. It claims a 26,000-gallon daily capacity.[33] Their in-field activity shows significant cost reduction and environmental protection.

Eagle Ford

Eagle Ford Shale Play

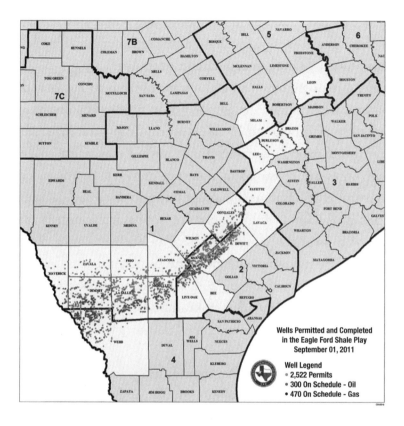

Wells Permitted and Completed
in the Eagle Ford Shale Play
September 01, 2011

Well Legend
○ 2,522 Permits
● 300 On Schedule - Oil
● 470 On Schedule - Gas

The Eagle Ford fields (Figure 12) are located in south central Texas, running in a long, slow arc from Dallas through San Antonio to the border. Reserves are estimated at 22 Tcf of gas and 3.4 billion barrels of oil (Bbbl). In 2008, it produced $3 billion in annual economic activity and hosted 12,000 new jobs. Those figures leaped to $25 billion and 47,000 jobs in 2011.[35] The local and state annual tax flows exceeded $500 million in 2011. Those funds, which pave roads, build schools, and pay salaries for inspectors and teachers, will only increase over the decade.

It is a new oil and gas play, with the emphasis on oil now, while gas prices remain depressed. These are deep wells, as much as 13,000 feet in the south, tapering up to 5,000 feet in the east. In 2011, 368 producing oil leases were reported, along with 550 gas leases. The shale actually emerges from below ground in the Dallas-Fort Worth metroplex.

The gas flows here can be enormous—up to 7.6 Mcf daily from one well. While the initial productions (IP) are huge, it drops off very quickly to 10 to 20 percent of initial runs, slowly tapering off over a relatively short period of five to 10 years. These flows are reported in a variety of ways. Estimated ultimate recovery (EUR) is an acceptable standard for accounting purposes. Reports of EUR can be very misleading. It is the result of a formula (simplified here as "AxB": proven reserves multiplied by recovery within a time period). It was developed by a federal agency, the USGS.[36] Their paper says quite clearly that this is a "guide." Guides are trail markers. Any hiker knows the danger of simply following trail markers without observing all other aspects of the terrain.

There are fields still producing, long after their expected lifetime. The industry colloquialism is "reserves growth."[37] There are gas plays that die out in a few weeks. It all depends on a sufficiently large number of factors, both known and unknown, that the concept is inappropriate, in the author's opinion. It is an accounting figure and should be treated as a mystery rather than a fact. IP, run rate and depletion figures are far more accurate. We shall visit this discussion in more detail in Chapter nine.

In the few years since 2008, it has become the center for Texas' E&P. Eagle Ford is unique in that the reserves for both oil and gas are extensive. The shale porosity allows fracing to function extremely efficiently. The 20 fields composing Eagle Ford will exceed total gross production of 207Mbbl in 2012, topping out at 1.4Bbbl by 2020. These 20 fields are just now being touted as the largest oil and gas discovery ever made in the U.S.

The nation's need for foreign oil imports will drop to zero in the same timeframe. The farther those imports have to travel, the greater the political risk involved. The lower the price of local oil, (West Texas Intermediate, WTI), the less valuable the imports become. The fears regarding the Caliphate of Europe may become real. They refuse to frac, thus enhancing their dependency on the Middle East fields. Political rescue can be preceded by economic dependency in any good novel. Only time will tell if Europe will stand on its own energy resources or fall before their neighbors' supplies. Is there any wonder why Texas avoided the recent recession, has added workers each year to its payrolls, and has no state income tax?

Marcellus

Marcellus Shale Map Appalachian Basin - USGS

The Marcellus shale (Figure 13) is perhaps the best known, certainly because of its proximity to the population centers of the East Coast. In early 2008, Terry Engelder, a geoscience professor at Pennsylvania State University, and Gary Lash, a geology professor at the State University of New York at Fredonia, surprised everyone with estimates that the Marcellus might contain more than 500 Tcf of natural gas (the equivalent of 8Bbbls of oil). Employing the same horizontal drilling and hydraulic fracturing methods from the Barnett shale, perhaps 10 percent of that gas (50 Tcf) might be recoverable. That volume of natural gas would be enough to supply the entire United States for about two years and has a wellhead value of about $1 trillion.[38] Approximately $11 billion in economic activity in Pennsylvania is directly attributable to the Marcellus, along with more than 214,000 jobs in the area. The field extends across West Virginia, New York, Ohio, Kentucky, Virginia, and parts of Tennessee. While New York has imposed a moratorium on shale development, the other states are moving forward with E&P. The website www.*fractrack.org* is a useful place to learn about current drilling activities and local issues.

Marcellus' development began soon after the Barnett's, using the same technology and skills. The well depths are mid-level, between 5,000 feet and 8,000 feet. Horizontal drilling is particularly successful here, as the shale allows for far larger extraction once penetrated. The volume flows here are regularly in excess of 1 million cubic feet (Mcf) daily. Landowners, initially reluctant to lease, are now fully participating in the production. Water use is an issue, and the state of Pennsylvania is encouraging drillers to use water from the old coalmines scattered throughout the area. Recycling, reuse, cleaning and substitution techniques are all growing in demand and application. Pennsylvania's Department of Energy regulates all activity.

Water use and disposal is as much of a concern in these Appalachian states as it is in Texas, albeit for different reasons. Proximity to populated areas and agricultural lands is the principal concern regarding hydraulic fracturing here. The State of Pennsylvania has imposed strict regulations on water use. In response, the water recycling firm in the Barnett shale, **Process Plants Corporation**, hopes to apply its recycling style to Western Pennsylvania frac sites. Oxygen is injected into the flowback and removes 97 percent or more of the metals present: aluminum, iron, and sulfur. Their reduction in water use encourages conservation. It also results in better communication between regulators, industry, and the local community.

Beneath the Marcellus lies the Utica field, far larger and costlier to exploit. Its size may be triple that of the Marcellus. Its reserve potential remains to be seen. As well flows decrease and technology expands, this field will be the next generation of shale gas development.

The Marcellus fields are the center of much controversy. "The End of Country" is a well-written commentary on one family's migration from opposition to acceptance of the hydraulic fracturing of wells on their property. By contrast, the new movie "Promised Land" exploits the fears of farmers and dairymen by suggesting that massive economic reward comes with a steep price—repudiating family roots. The true story in Western Pennsylvania is at once more congenial and more cantankerous. The drillers have to prove themselves before they are accepted. Negotiations are hardly one-sided. Landsmen have lost more than one price fight for lease rights development to smart country folk. The ultimate reward is financial, certainly, but it also includes an abiding respect for the drillers. Most folks who live in the rural parts of the state have their own way; they are more country than Nashville. The rebellious nature of the backwoods Pennsylvanian remains intact. Whether facts or fear drive the controversy, time will tell. It is never a wise bet to short change the local community: penny wise, pound foolish.

Bakken

Bakken Type Log

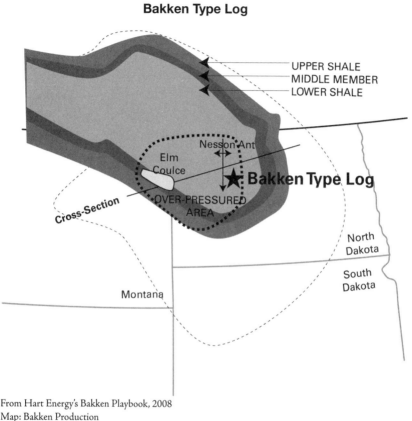

UPPER SHALE
MIDDLE MEMBER
LOWER SHALE

Nesson Ant
Elm Coulce
★ Bakken Type Log
Cross-Section
OVER-PRESSURED AREA

North Dakota
South Dakota
Montana

From Hart Energy's Bakken Playbook, 2008
Map: Bakken Production
July 24, 2012

The Bakken fields (Figure 14) are in North Dakota, Montana, and Saskatchewan, Canada. Originally estimated to hold three to four Bbbl of oil, recent recoverable estimates exceed 24 Bbbl.[39] The figure for recoverable gas is 1.4 Tcf., although this is essentially an oil play for the foreseeable future. These recoverable figures remain moving targets, as both technology and access evolve. Today, the 500,000-plus barrels per day (BOPD) produced far exceed transportation capabilities to ship to refineries.[40]

Local pipelines and mid-sized refineries are being built by the regional tribes to accept the product. The average salary for an oil field worker in North Dakota in 2011 was $84,000. Rail remains the primary transport vehicle; hence, Mr. Buffet's acquisition of Burlington Northern.

The Canadian resource has been described as the largest find in that nation's history. These wells are the most expensive to drill in both the U.S. and Canada, at $8 million to $12 million per well. There are today more than 3,641 wells working the Bakken fields. These deposits are twice the size of Alaska's North Slope, with far better recovery rates at a much lower cost. As a result, North Dakota has surpassed Alaska as the nation's second largest oil production state, having passed the downward-sliding California last year. North Dakota is well on the way to 1 Mbbl of daily production.

The **takeaway capacity** must grow, simply to keep pace with production. Delivery is the real challenge. Warren Buffet bought out the railroad simply to "corner the market" in oil transportation: **Berkshire Northern Railway** (BNSF). Sand is brought in from the Alberta tar sands disposal, and oil is sent off to Canada. **Canadian Pacific's** shipping has increased by 90 percent year over year. Still, production increases beyond delivery capabilities.[41] Terminal capacity in North Dakota is nearing 300,000 BPD and will expand to 750,000 BPD by the end of 2013. Capacity and daily use differ markedly, with daily use as of July 2012 at 70,000 BPD.[42] Railroads tout their specific site delivery capability verses pipeline, which is point to point. Both types of delivery are clearly needed.

Trucking provides jobs but also destroys roadways. A new refinery has been approved in North Dakota to convert oil into diesel and gasoline. While small by industry standards, building it close to the energy source makes sense.[43] Pipeline construction continues, as production grows. Tulsa, Oklahoma-based **ONEOK** is contracting to build a 60,000 BPD line, with additional pumping potential to bring it up to 110,000 BPD. The cost is $500 million, or nearly $1 million per mile for the 525-mile-long line. Three NGL processing plants are part of the major facility, with plans to come on line by 2014.[44] The Canadian Bakken is also in regulatory approval stage for a 100-mile-long pipeline to carry oil from Montana and Saskatchewan to eastern Canada, and to the Texas refineries on the Gulf of Mexico.

Meanwhile, the local economic results are clear: 65,000 new jobs, $12 billion in new economic activity since 2010, and $1 billion in new tax revenue to North Dakota. The state has a balanced budget, a surplus, fully funded pensions, and has reduced taxes across the board. The benefits to the nation will be discussed later, but each of these fields reduces dependence on imported oil by their production equivalent. The arithmetic is simple: 500,000 bbls daily from Bakken equals 500,000 bbls reduced imports.

Niobrara Shale

Niobrara Shale Project

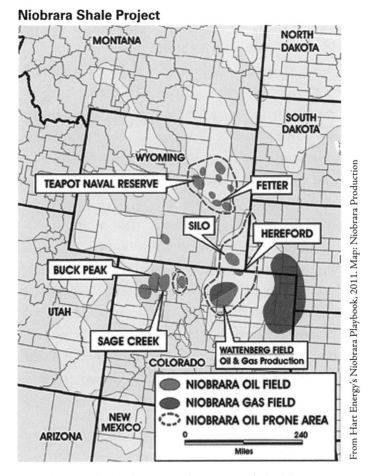

From Hart Energy's Niobrara Playbook, 2011. Map: Niobrara Production

The Niobrara shale deposits (Figure 15) hold an estimated three to five Bbbl of hydrocarbons, 90 percent of which are oil and NGLs. The first well was drilled in 1876; today, the region has finally emerged as a potential top-line shale field. Its hydraulic fracturing and horizontal drilling exploration and production only began in 2009. The hundred or so wells completed are from 3,000 feet to 7,000 feet deep. Their completion costs range from $5 million to $6 million per well.

Colorado, Nebraska, and Wyoming sit atop the formation, with the majority in Colorado. Today, the oil and gas industry as a whole employs seven percent of Colorado's workforce, for a total of 137,000 jobs. Of the 40,000 wells in service in the state, virtually all are verticals. While the few new horizontal wells are initially very productive at over 1,500 BOPD, the extraction rate quickly declines to about 12 to 15 percent of that figure.[45] Better extraction techniques are developing in the field, but these figures can be disappointing to a landowner hoping to cash in on mineral rights today. These rights are simply not worth what they are in North Dakota or Pennsylvania. The process is too costly to warrant significant premiums on land leases.

The variances between the sections of the shale field are significant: Wattenberg flows nicely, while faults, drill depths, frac radii and gas-to-oil ratios, or GORs, have yet to favor E&P in Silo, Wyoming. Higher GORs indicate more gas and less oil. Typically, oil is more profitable to extract than gas over shorter time lines. Getting capital back out of each well is a financial driver for every company in the field, small to large. At these prices for a completion, and the current extractions rates for a well, the challenge remains pricing.

Increasing the number of frac stages will help the drilling completion and EUR (estimated ultimate recovery). From eight to 15 initial stages in the Bakken, these have steadily increased to 20 to 40 today. Applying this advanced technology to the Niobrara should open up EUR significantly. Every well is different, with varying deep terrain in eastern Colorado and southern Wyoming. There is little that is "normal" about these fields. Fracing technology evolves within each well and across wellheads,[46] and ultimately, demand has to catch up to supply for these fields to play successfully.

Haynesville

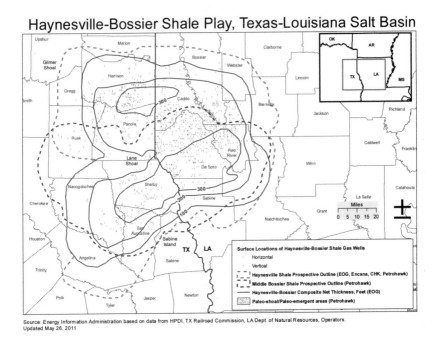

Haynesville-Bossier Shale Play, Texas-Louisiana Salt Basin

Source: Energy Information Administration based on data from HPDI, TX Railroad Commission, LA Dept. of Natural Resources, Operators.
Updated May 26, 2011

A deep pocket of very rich gas deposits sits below the Texas/Arkansas/Louisiana borders. The Haynesville (Figure 16) may have 240 Tcf of gas.[47] EUR is more than twice that figure (caveat emptor). Another field, the Bossier, overlays the Haynesville, yet both have depths that make development costly. The wells are 10,000 feet deep and more, the fields are rich with gas, and the flows are strong. Economic benefits as estimated in Louisiana are 100,000 jobs with an average field worker salary of $80,264; $1.4 billion in tax revenues; and $40 billion in gross economic effect between 2010 and 2012.[48] Today, 1,860 wells are producing a record amount of gas.

The depth is a challenge. The heat and pressure at these depths are both an obstacle and an opportunity. The pressure does force more gas upbore, more quickly than in shallower plays, but the initial production (IP) drops substantially. Each well may produce 2 Mcf IP with five to seven Bcf in EUR. Recovery rates are measured in months rather than years. Prep and perf in 10 to 15 stages of 50 to 60 feet are common in the Haynesville fields, with proppants designed to the hole. Higher gel mix and larger mesh sand with ceramic or resin coatings tend to be more successful. Flowback, scaling, and corrosion issues abound as a direct result of the temperatures and pressures found downhole. Embedding into the soft shale often reduces the proppant's useful life.

Today's depressed natural gas prices have significantly reduced the number of E&P rigs drilling, while production rigs are stable. The fields are rich in both gas and oil. The price of oil drives current E&P, and the price tag of $9 million to $10 million for a completion makes further exploration uneconomic for now.

Despite the price-directed E&P declines, Haynesville surpassed Barnett as the nation's leading shale gas play as of March 2011. The fields have the top five producing shale gas wells on the planet as of May 2011.[49] **Chesapeake Energy** dominates it; however, their recent senior management turmoil may have distracted the firm from details of operations. The CEO, who has since stepped down, recently stated that the firm overpaid by a factor of two for leasing rights in Haynesville.[50] Smaller firms are taking up the slack and producing in fair quantities.[51] Haynesville may see production declines over the next decade, depending on accessibility of the resource, capital funding, and gas prices.

Fayetteville Field

Fayetteville Shale Play, Arkoma Basin, Arkansas

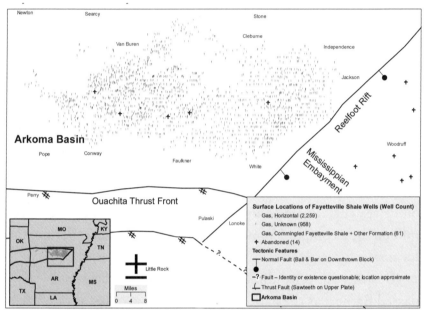

Source: U.S. Energy Information Administration based on data from HPDI, Arkansas Geological Survey, Ratchford et al (2006)
Updated: May 31, 2011

The Fayetteville field (Figure 17) was one of the first to be developed after Barnett proved in 2005 the efficacy of hydraulic fracturing and horizontal drilling. Well depths here are shallow, from 1,500 to 7,000 feet. Technically, recoverable natural gas reserves are estimated at 20 Tcf.[54] Temperature and pressure are both lower, resulting in lower IP, but softer decline rates at 56 percent. Costs are lower as a result, at about $2.8 million per well. A publicly traded company in the state reported production rates of nearly 2 Mcf daily.[55] The average well may have a EUR of 1.7 Bcf or more.

The relatively long history of the development allows some projections to be offered. [56] Well life can be 40 years as the recovery rates are lower, slower, and more consistent. These will create very significant wealth for the counties of north central Arkansas in the form of royalties: $1.2 billion has been paid since 2005.[57]

This field, along with Haynesville and the Barnett fields, accounts for a sizeable minority of U.S. shale gas production; together, the three accounted for 25 percent of U.S. annual production in 2011.[58] The economic impact to the economy of Arkansas has been estimated at $20 billion between 2006 and 2012.[59] In 2012, 16,000 jobs are being added, with total economic impact of $4 billion.[60] The estimate for EUR is as high as 20 Tcf. Average pay in the area is $74,500, a 24 percent increase for the counties involved since 2006; this, despite the recession and the loss of much of the manufacturing jobs base. In 2011, 94.4 Bcf of gas was extracted. Between 2008 and 2011, $2 billion in tax revenues were added to local and state coffers as a direct result of the drilling operations for shale gas.[61] The rural counties of northern Arkansas have benefited greatly.

E&P rig counts are down to little more than a dozen for 2012, while completions are high at nearly 2,000.[62] "Long term" is used today for drilling prospects in the Fayetteville area of north central Arkansas. Until gas prices reach breakeven for the producers at $6 to $7, they will step back from new drilling. As evidence, Chesapeake has just sold $5 billion worth of leases in the area to other firms, including **BHP Billiton**; their current capital needs are dictating the focus on short-term vs. long-term capital expenditures, or "capex."

Utica Shale

Utica Shale Map

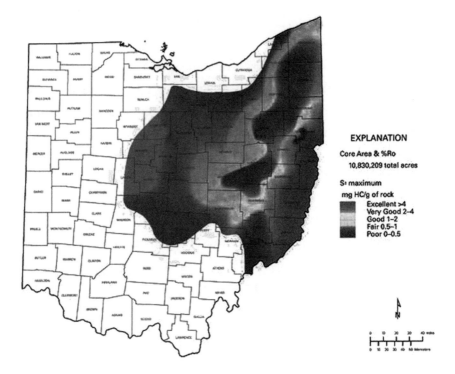

EXPLANATION

Core Area & %Ro
10,830,209 total acres

S₁ maximum
mg HC/g of rock

Excellent >4
Very Good 2–4
Good 1–2
Fair 0.5–1
Poor 0–0.5

Beneath the Marcellus shale, deep beneath it in most places, lies the Utica shale deposit (Figure 18). At depths ranging from 3,000 feet to 7,000, this is a deep drill, so its full exploitation may not come for decades. Its high-carbonate content makes it similar to the Eagle Ford shale—brittle and more challenging to frac effectively.

The Utica deposits rise closer to the surface near the provincial borders of Ontario and Quebec, and in Eastern Ohio. In Canada, it is being developed commercially by **Questerre**; in Ohio by

Chesapeake. It has potential recoverable reserves of 1.5 Bbbl to 5.5 Bbbl of oil and 4 Tcf to 16 Tcf of natural gas. These figures are very early estimates, and some figures come in as high as 60 Tcf.[64] NGL potentials have yet to be accurately measured. These are massive figures. Recall that today's total potential recoverable gas reserves for the U.S. are at 1,750 Tcf.

The few wells flowing today are reported, in press releases from **Chesapeake**, to be in excess of 5 Mcf daily. **Questerre** reported that similar initial runs declined, as expected, to 1.4 Mcf daily after five months.[65] While a few hundred permits are open, less than a hundred wells are in operation. Lease sales are very active, with $15,000 an acre being common. Royalties are growing at a slower rate. The difficulties in reaching and extracting these reserves may be offset by future technological advances for high-carbonate shale, as well as the in-place availability of existing pads, roads, and processing facilities. Its thickness, in excess of 1,000 feet in many areas, suggests fracturing technology must expand its range of effectiveness by several factors for horizontal drilling to be most effective.

The economics of the play in Ohio for 2011 weigh in at nearly $1.7 billion gross production, with more than 13,000 jobs and $33 million in tax revenues. While still small, these figures have grown from a nominal base in three years. By 2015, Ohio estimates 200,000 new jobs and $22 billion in increased economic output.

Two firms, **Hilcorp** and **NiSource**, have recently announced plans for a joint venture to exploit the Pennsylvania Utica shale. In Ohio, 92 wells have been drilled, from 285 permits. This is very recent developmental activity, although Ohio has been active in fracturing since the 1950s and has 80,000 hydraulically fractured wells.

Woodford

Woodford - Kulkarni Shale Map

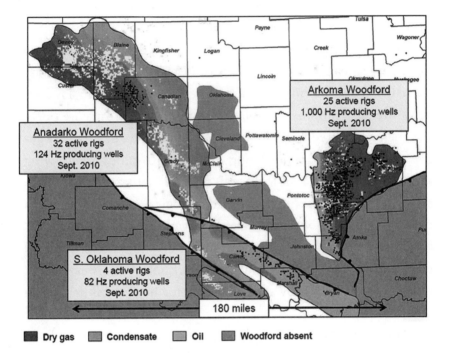

As shown in Figure 19, the central and southern parts of Oklahoma have significant oil and gas deposits, with technically recoverable amounts estimated today at 4 Tcf of gas and 250 Bbbl of oil. With 6,000 to 16,000-foot wells, these are the deepest shale wells globally. Each well cost $2 million to $4 million to drill. Their production runs of 10 Mcf daily for gas, and 740 bbl per day of oil, allow for capital recovery in less than a year,[67] at today's prices. This same report from 2010 shows what is obvious today: Horizontal, multi-string holes are extremely economical, despite their significantly higher costs. The State of Oklahoma earns 37 percent of its revenue from oil service taxes, or nearly $400 million. Jobs have grown, while the average income of an oil field worker is 50 percent higher than 10 years ago.

This is one of the earlier shale plays, given that they have been drilling here since 1934. Extraction requires 3-D seismic analyses to determine precisely where to drill. In the glow of an East Texas dawn, the rock is folded upon itself in many places, but for all its tricky layering, it can also be ideal for fracing with the right tools. The brittle, silica-rich rock formations are susceptible to ultra-lightweight proppants. Rather than 80, the propps are graded down to 200, which is nearly as fine as talcum powder. Because the rock is, in a sense, pre-fraced, once the bore is perforated, or "perfed," during the fracing process, the flows are enormous. This technology is integral to proper drilling access. Because much of this is proprietary, only a few E&P operators and OFS firms know the secrets of accessibility. This is state-of-the-art frac science in the fields of Oklahoma.[68]

FRACING: HYDRAULIC
FRACTURING

Hydraulic fracturing is very much a necessary
part of the future of natural gas.
—Ken Salazar, U.S. Secretary of the Interior

racing is the process of opening seams in deep rock with high-pressure fluids mixed with sand and trace amounts of chemicals. The liquid is extracted, leaving the sand in place. The sand holds open the fractures in the rock. The hydrocarbon then flows out along these fractures, up the borehole, and is transshipped.

This very simple description of a very complex process is the basis for this book's title. Fracing is but one step in a long chain of events that results in electricity for your home, gas for buses and heating, chemicals for the plastics scattered throughout your house (computer monitors, dog dishes, cell phone cases, sippy cups), as well as the important, little things in life: WD-40 and duct tape and barbeques in July.

The story of the development of the frac process is one of perseverance. Hardheaded men with engineering backgrounds and some knowledge of geology have been poking holes in the hard-scrabble rock of West Texas and Oklahoma for decades. Many have failed, many times. The Barnett has been the toughest to crack. Wildcatters have spent blood and treasure for decades trying to break through an impenetrable mass of overlying rock to the expected reward below.

As we learned in Chapter 2, fracturing of underlying rock formations for the extraction of petroleum, natural gas, and NGLs dates back to the mid-19th Century. By 1949, a total of 332 wells were "hydrofraced," with a production increase of 75 percent. News of the increased productivity spread, and by the mid-1950s, some 3,000 wells were being fraced.69

The fluids used today are water, a gel with the viscosity of dishwashing detergent, a propane gel, an acid wash, or a combination of ingredients called slick water. These have developed over the past four decades, and today, ultraclean gel agents and "encapsulated breaker systems" enhance the fracs' ability to remain open.

Three main types of fracing are pursued today: slick water, gel, and acid.

Slick water is most commonly used in deep holes, where the water is under extreme pressure. It is this combination of water, gel sand, and chemicals that is often referred to when speaking of fracing. Slick water can substantially increase the pressure downhole, encouraging a wider frac radius for each stage. The wider the radius, the greater the production potential. The chemicals used are friction-reducing agents, biocides, scale inhibitors, and surfactants.

Friction-reducing agents (polyacrylamides) speed the liquid flow. Biocides such as bromide or methanol prevent biological growths from gumming the mixture (yes, there is a wide assortment of life three miles down). Scale inhibitors are hydrochloric acid and ethylene glycol, and do exactly as their name implies.

Surfactants promote the "wetting" of a surface by increasing a liquid's ability to disperse or emulsify (think Dawn dishwashing liquid). They reduce the surface tension of the liquid, which is critical when the liquid is under extreme pressure, down a very deep hole, forcing open rock fissures millions of years in formation. These can be butanol or ethylene glycol monobutol or a host of propriety products.

Gel fracing was the first approach to hydraulic fracturing. It began as Ferris bombs—they were called torpedoes—that were dropped down the early wellbores. This quickly evolved into dynamite, then napalm laden with sand. Hydrochloric acid was tried, as a disposal into an old well. It released more oil. These "natural experiments" evolved with the industry over the hundred years between Drake's first well in Pennsylvania, Halliburton's gels in the 1930s, and Ferris' experiments, through Mitchells' accidental breakthrough. Gel fracs evolved with the times.

Today's **Gelfrac** is on the cutting edge of the frac world. It is waterless, as it uses LPG, or liquefied petroleum gas. This is propane- and butane-based. As propane is an NGL lifted from wells, the recycling event is obvious. **Gelfrac** is the Alberta, Canada-based firm that has developed the technology. While the cost is 20 to 25 percent higher than slick water, the petroleum product recovery rates are also higher—by the same amounts. The system employed is a closed system, which recovers all of the gel, promising higher, quicker, production rates. Its use reduces the environmental impact of cleanup and may reduce truck traffic by as much as 90 percent. In drought areas, as well as populated lands where water pollution remains a concern, this technology is quite promising. Slick-water recycling rates are about 20 percent. Despite these recycling efforts, a typical 4-million-gallon frac consumes more than 800,000 gallons of surface water.

Safety concerns are addressed by the complete computerization of the closed-system process, with field workers well-removed from the site. Training in the application is required, and is evolving. Flaring is significantly reduced, which eliminates carbon emissions concerns at each well site. The gel remains downhole, dissolved into the gas, awaiting extraction and reuse by the energy company.

Acid fracing is an older technology that is used where the rock is susceptible to the etching of an acid wash. This is typical in rock with low permeability, such as limestone and dolomite. Pumping the hydrochloric acid downhole forces it into the rock formation where it etches channels. The hydrocarbons then run through these traces and upbore.[70] The challenge is sufficient to have reduced its application to these types of sedimentary rock formations and workovers. These are the restoration, prolonging, or enhancing of production from older wells. Acid fracs often do not use proppants to hold open the etched channel lines. This results in "wormholes," or loss of product laterally back into the rock, unless chemical additives are utilized to close these rock capillaries.

Proppants

The proppant for these frac jobs is essentially sand, often riverine. While size matters to the fracer, 85 percent of propp sand today is -20 to +40 mesh size. Its concentration in the water base has increased markedly with gel stabilizer types. Today, it averages seven pounds per gallon. During the course of the application, this may begin at two pounds and increase to 20-plus pounds by the end. Resin coatings are often added to the sand to enhance their tensile strength—their ability to withstand the pressure of several thousand feet of rock overhead—and to maintain this tensile strength though the life of the completion.

Other propps may be a wide variety of substances, graded for sphericity, as Jake told us. As an example, high tensile-strength bauxite can be rolled into balls to act as the support against the high pressure of the rock above. It can also be designed in an angular shape and mixed with the balls; this shape prevents flow back as the high-pressure water is extracted, leaving more of the formation-opening bauxite in place. Synthetics, ceramics, and resin coatings are among the choices available to the fracer. Each choice depends on the engineering of the hole, the rock below, the skill and function of the men and equipment—and a goodly dose of luck. As riggers will tell you, "Each hole is different; each has a lesson to teach."

There are 50 proppant manufacturers globally. The demand in the U.S. alone has increased from five billion pounds in 1999 to 28 billion pounds in 2011. Propp size has gradually reduced as water pressure increases.[71] The manufacturer today has the demands of quality, quantity, and sphericity to offer to the consuming rigger. **Momentive Oilfield Technology**, a major supplier of coated proppants, manufactures to the engineering designs of each hole, based on its proprietary standards.[72]

Frac spreads (a combination of four frac units: a blender, a sand truck and the ancillary equipment in support) have multiplied significantly over the past decade, in number and in size. These units provide the power and the proppants for the frac. The typical horizontal borehole today may run 10 to 40 stages, or fracture events, downhole. The complexity of such an operation demands the highest quality people and equipment. Fortunately, this is readily available in the States. As the center of E&P for more than a century, we have the experience, equipment, knowledge, and technology to drill deeper and more productively than anywhere else in the world. The competition among hundreds of small firms, each testing new ideas and improving on existing ones, makes for a very lively (and often secretive) marketplace. Such a market allows the best—and the best capitalized—to rise quickly. Capital intensity can be a barrier to entry for newer global firms entering such a crowded arena.

On a global scale, the U.S. has, for now, the leading edge in technology, manpower, and skills to drive the industry forward. The global demand for energy continues to grow, despite the systemic economic challenges of the past five years. Demand in Asia, India, and Africa must be met.

Setting the Stage

The U.S.-based demand for these combinations of experienced men and equipment can mean a delivery backlog of nine months or more today. Their components and functions have grown with the demand. Fracs of 10,000 horsepower are mid-sized jobs today, as horizontal boreholes several miles in length, and several radii in composition, can require much more pressure, liquids, sand, and proppants. Fitting 16 to 24 fracs on a hole is becoming easier. The economics drives the capital: Cost recovery can occur in less than a year in some cases. The environmental footprint shrinks as well.

These are labor-intensive, experientially based groups of men and equipment. The professional rigger today needs to know the basics of the hydraulic fracturing process, stress issues, fracture geometry, controls on generated length and width, fluid and proppant selection, quality control, quality assurance, and personnel safety. The team leaders have to know the use of dynamic data, wellbore radius, effective fracture half-length, equivalent skin, and the relationships among these. The field computers must be able to simulate and field test basic analysis procedures as well as track events and characteristics down hole in real time. Ultimately, the team wants to predict long-term rate performance and recoverable volumes. Accuracy here is always a moving target. As prices change, so does EUR. As depletion rates (DR) increase, demand for drilling does so, as well.

While these men drive the industry, robotics may be replacing them at the cutting edge of frac tech. Lessons from NASA, the railroads, international robotics, and the nuclear industries are being applied to drilling and completion technology at a swift pace. The field computer today can view and track pressure, frac radius, porosity, permeability, hydraulics, and a number of other technical aspects of the well. This information is available in real time. You can "see" the hole at 50-foot intervals along its entire length while monitoring conditions of the pipe, casing, fluids, and surrounding rock. As hard as this rock is, and as deep as these wells are, they still are constantly changing. One of the extraordinary changes in well technology has been this evolution of deep-bore knowledge.

The cutting edge of research is also here, at the simplest level. How does the sand work in the hole? What choices does the driller have for shape, size, strength, porosity, permeability, and viscosity? Permeability, stress, and fluid loss are three of the key issues that ongoing research addresses today. Virtually all sand is naturally produced. These are preferred in the Bakken, Eagle Ford, and Permian basins. Manufactured proppant such as ceramics, bauxite, and plastics (resin) are used in the deep bores of Marcellus, Haynesville, Bakken, and Fayetteville.[73]

A typical deep frac may require one to five million pounds of sand. This is delivered by rail initially, so a job would require five to 25 railroad cars. Trucks then haul the sand the "last mile" to the well site. You can begin to see the challenges to manpower, vehicular demand, traffic management, road erosion, emissions, and fueling—for just one well!

The seemingly simple process of sending sand and saltwater down a hole is far from simple. The choice of sand type, its source, and its composition varies with each wellbore. Northern sand, Chinese sand, and manufactured proppants are each significant choices. Which type of proppant for each hole depends on a number of factors including depth, pressure, porosity and permeability eight to 12 miles away from you at the end of an 11-inch-diameter pipe weighing more than a hundred tons. How long, deep, and frequent will be the stages? What are you going to pull up: oil, gas, or NGLs? Who is doing the work, what is the time frame, and how do regulations impact what you put downhole and what you bring back up? How and where do you deploy the waste? What is the life cycle cost?

U.S. Silica, the preeminent proppant supplier in North America, has been "doing sand" for more than a century. They design the sand propp to five essential aspects: low turbidity, grain-size consistency, high crush resistance, high sphericity, and low solubility. Mesh sizes (permeability of the proppant) vary with the frac, depth, temperature, flow porosity, and viscosity of the liquid. Ceramic coatings can be applied to the sand, plastics, or metal (bauxite or aluminum) proppants. Produced by **Saint-Gobain**, out of Arkansas, **sintered bauxite** is a ceramic propp containing corundum, an extremely hard substance used to grind the glass for telescopes. This can reduce or inhibit the crush rate of the proppant. In deep wells, size matters, and this proppant can make all the difference in each frac stage. Smaller-sized proppants penetrate farther without closing the frac vein, or "bridging." Lighter propps can retain more shape (fracture width) under extreme pressure when crushed. This is critical, as pressure at 12,000-foot-plus depths can exceed 10,000 psi.

The amount of proppant soluble in the fluid differs with weight, or mesh size. Its presence in the liquid is a function of mass, as sold. Greater solubility allows more and finer sand deeper into the wellbore. Coarser sand allows for greater upbore flow. As long as its use precludes excessive pressure, coarser sand can be far more economical to push down bore, to place, and ultimately to extract the hydrocarbon. Simply put, it leaves larger fractures. Deeper wells must turn to other choices, such as resin or ceramic coatings. The choice depends on the well, the rock, and the depth.

The modeling of the fluid dynamics of sand-infused liquids is an ongoing aspect of deep research in frac tech. Understanding this process of "packing-the-well" annulus (Latin for little ring) requires the application of advanced fluid mechanic mathematics.

After a few pages, we have only touched the surface of this discussion. Choice of sand is far more involved than you would think when first we began discussing proppant. It's not just sand.

Today, science and the applied technology are shaping the world of hydrofracing. To quote the summary from a textbook, "Recent Advances in Hydraulic Fracturing":[74]

> …pretreatment formation evaluation, rock mechanics and fracture geometry, 2D and 3D fracture-propagation models, propping agents and fracture conductivity, fracturing fluids and additives, fluid leakoff, flow behavior, proppant transport, treatment design, well completions, field implementation, fracturing-pressure analysis, post fracture formation evaluation, fracture azimuth and geometry determination are domain knowledge requirements.

If this passage confuses rather than enlightens you, then the point is made. Remarkable men like Jake and the seemingly unschooled men who report to him are far more knowledgeable than appearance may indicate. Each of these "domain knowledge requirements" must be on the knowledge circuit board of an effective rigging crew. Today, the crew must have working experience of these models, behaviors, and technology approaches. Those who do, or who learn on the job, rise quickly through the ranks. These young men will be the capitalists of the next three decades, the leaders in the industry.

And they are predominantly men. There are women in the industry, but they must work even harder to prove themselves. That and other aspects of the job are a deterrent, as is evident in a chance encounter I had with Gina, a former oil field worker. Gina is tall and strong. Hers is a grit-to-glamor story: A single mother of two teenage girls, she worked in the fields for nearly a year before retiring to become a hair stylist. As a roustabout, she performed all the hardest labor in a camp: bucking, running chain, reaching from the monkey board, pulling tong for the toolpusher. "Hell, for $24 an hour how could I refuse the job? They were hiring for their minority quota and they needed women. They only got one, but they did get me."

The men hardly took to her. She was tested every day, just because she was a rare bird in a man's world. They tried to trick her on safety, without going so far as to hurt her. A field manager tried to proposition her. When she made it clear that she was there to work—and his was not the sort of "drilling" she got paid for—he made her job all the more difficult. She persevered. Slowly, respect grew for her skills – and for her strength. "You need to rely on every person on the crew, even if you think she is gonna fail." So, grudgingly, the men accepted her.

However, when offered the job at the hair salon, she grabbed it without hesitation. The pay was the same, but the hassles disappeared. Gina was a rare find in a man's world, a tough lady roustabout.

Bidding a fond farewell to Gina and getting back to the business at hand, we have thus far set the stage for the frac by describing what it is, what is used in a frac, and some of the technology; however, all this is of no use without a hole to frac. Once we have the hole—the wellbore—we have to prepare it for the frac. Let's spend some time on the drilling and prep process.

Drilling Preparations and Operations

First, you must find the rock. It is buried deep beneath your feet. You use the eyes of robots, of computers, of large thumping machines. Your search is for a sponge-like rock, an absorbent material of great porosity, but little permeability. The shells and cells of countless trillions of small creatures have drifted down to the "once and future" sea bottom. These antediluvian seas, where today great cities, farms, and schools rise, are the deep reservoirs you seek. They have risen, fallen, dried, and have been covered over and forgotten. These deposits are crushed by time, temperature, and pressure a mile or more beneath the surface today. These small granules even today are slowly coalescing into the coal and oil and gas of tens of millions of years from now. You want it today—this gas, this bitumen, this early oil.

You look for sedimentary rock. You look for cap rock that covers these deposits, holding them in place. For years you could not penetrate this shale cap. George Mitchell showed you how. Now you can get beneath the capstone to the rich deposits below, and you can exploit it quicker, deeper, and more efficiently than ever thought possible. Seismic analysis allows you to image these traps, sinks, domes, faults, and pinchouts—all the little nooks and crannies—for the richness they hold. The geologist has described the stone. The engineer has imaged it in 2-D and 3-D, and now 4-D. The Cray supercomputers developed for the defense industry sort these data mines for the long runs of a reservoir. **Texas Instruments** pioneered the seismic imaging process. Today, arrays of computers, like artillery ranks, locate the gas and oil in exquisite detail. You still have to guess, of course. In bass fishing, when you equip your boat with a fish finder, your chances improve significantly. So too, when fishing for hydrocarbons. The game has come to you. Gravimetric, magnetic, radar, and seismic tools give the petrochemical engineer and the geologist and the geophysicist the advantage.

You need to drill test wells to demonstrate presence, pressure, production, and permeability. The first well is exploratory—**wildcat wells.** In the past, these were harder than a Vegas bet on Saturday night. Today, most wildcats are near-perfect; the hit ratio can exceed 99 percent. They remain tight-hole operations—confidential and proprietary. **Thumper trucks,** with massive hammers creating long deep sonic shock waves, are then brought in to determine the boundaries of the field. They also determine the well's viability by measuring the bounce-back echo.

Appraisal wells surround a good well to determine field size and potential. Today, **3-D technology** has taken the place of these definition wells. **Development wells** are the production workhorses of the field. If this is a new field, these wells cost $500,000 to $6 million each. It all depends on the depth, the rock, the length of horizontal run, and the number of holes punched through the pad. Sixteen bores can go downhole from one pad today, arraying out into a deep field of hydrocarbon deposits. If this is a recompletion well, the cost is less: These may be two miles down. These horizontal wells can have 15,000-foot centers and 7,000-foot laterals over 160-acre plots looking for a four percent recovery rate.

This means they can pull up perhaps four percent of the hydrocarbon in the field.

Let's step back in time for just a moment. Before you drill, you have to acquire ownership or access to the land surface. You gather as many contiguous land leases as possible. Your **landsmen** (Matt Damon in "Promised Land") go to the landowners—ranchers, farmers, homeowners, businessmen, Native Americans, and anyone else with clear title. Their task is to negotiate the deal. These landsmen are slicker than a New York raincoat in April. Watch your back. If you saw "Promised Land," you saw one in action. The negotiations can change with each landowner, with the time of day. If you are a potential leaseholder, seek competent local legal aid. Don't do this on your own. Time usually works to your advantage. The Chinese have an expression: "Waiting is…" Take your time. Review the contract with a knowledgeable attorney who specializes in this area. Make changes as you see fit. Consult with your local city or county officials, too. Their input may be valuable. Always sit with your back to the wall. Landsmen will cut the best deal possible—for their employer. They aren't necessarily crooked or shady; in fact, few are. But they have a job to do. They negotiate the best price possible, period. You had better do the same.

The **lease** is real property, and can be traded or sold. The lease is a legal document that grants the right to explore, drill, and produce. In exchange for a **royalty**, and perhaps a bonus up front, the lease has bought a limited time during which it can operate, usually one to five years for the preliminary term. If hydrocarbons are found in numbers capable of future production, a secondary term is extended. **Overriding royalties** often flow to others: the landsmen, geologist, promoter, and so on.

The, oil field service firm (OFS) has been hired to punch the hole. These firms own or lease the rigs, and they set the **spud date**, the day the drilling is to begin. They have set up the Authority for Expenditure or estimated cost pro forma. Joint operating agreements have been agreed to, and responsibilities defined. The **plat**, or land use document, at this stage is prepared and registered.

The land is then cleared for the rigging operator and the equipment. A thick **mat** covers the area to promote proper drainage and ensure any spillage is controlled. The **pad**, from which the work is done, may have another, stronger mat placed just beneath its footprint. The **reserve pit** is dug, lined, and approved. This will hold the drilling mud, cuttings, and fill. Water can be a challenge or quite easy, depending on your locale, regulations, and the community. You either truck water in, dig a well, or lay in a pipeline. Today, you may be using recovered water from a previous drill, as you attempt to recycle your water. You may bring in water from old coalmines, if you are in the Appalachians.

Deep wells will have a **cellar**, just beneath the pad, to hold the **blowout preventers** (BOPs). If this well is shallow or vertical, the rig brought in will be truck-mounted. Deeper or horizontal wells require far greater support for the entire drill string, and are built on top of the pad. These **derricks** are the size of 10-story buildings. They are the oil well icon conjured by the city dweller. The trucks used are massive. They negotiate access roads with difficulty. They tear up roadbed. They destroy local habitat and crops. Today the drilling firm often contracts to fully restore habitat, often to better-than-original condition.

You have determined the exact angle of attack for the hydrocarbon reservoir beneath your feet: its depth, porosity, and permeability. You have filed all documents with the state and local county regulators, including land use, chemicals list, worker status, equipment roster, road site management, and final approvals and permits.

Spudding begins with a **conductor hole**, a large-diameter hole that provides stability for the rig pad and isolation from any surrounding water source typically to a depth of at least 800 feet. The well bore begins in earnest once the **conductor casing** is laid in. Power, hoisting, rotation, and circulation systems are engineered for each hole. The power trains are usually combinations of single, duals, and quads: 1,000 to 4,000 horsepower diesel engines. Their AC is converted to DC for the drilling operation, the hoisting, rotating, and circulating units. Diesel fuel tanks are stored on their own **control mats**.

The **mast**, on a portable rig, is the most visible element of the rig. It supports both the **crown block** at the top of the drill string and the entire drill pipe itself. If it is assembled onsite, it is a derrick. Most today are treble derricks, so named because they can hold aloft three vertical lengths of drill pipe, each between 30 and 45 feet in length. Treble derricks can soar to a height of nearly 200 feet. They are rated to withstand hurricane force winds and may have hoisting loads of more than a million pounds. The **hoisting line** is a thick braided wire rope spooled around a revolving drum called the **drawworks**. The operator controls the motion up and down and its speed with these geared works.

This drilling line, the hoisting line, goes over and through the crown block at the top of the derrick several times, then is attached to the **traveling block**, from which hangs a massive hook via a swivel that allows the drill string to rotate. This in turn is attached to the **Kelly,** a 40- to 54-foot-long strong steel multisided pipe that can be gripped by the horizontal rotary table. This Kelly holds the entire drill string and turns it in rotation. This drill string can be miles long and weigh hundreds of tons. The Kelly and the **rotary table**, supported on the floor of the rig, support and rotate the entire length of drill pipe.

Rated for various hole types, the drill pipe sections can be 30 to 45 feet in length and 2½ to 11½ inches in diameter. They are threaded on each end; each new pipe section is screwed, or rotated, on to the last pipe length after a pipe dope is applied. The **tool joint**—the threading—has been welded onto the pipe and is tapered for ease of connection. Each **joint,** or pipe length, is reused after each wellbore and graded for further reuse. They are stacked next to the rig and loaded from the racks through the V-door in the derrick. Often they are pre-stacked next to the derrick, ready for connection by the derrick man on the **monkey board,** the small platform about halfway up the derrick. This is one of the most dangerous jobs on the rig—unless you count every other job performed on the site, 24 hours a day, seven days a week. The repetitive nature of the work is as dangerous as the work itself.

The driller rotates the new joint for most of its length, down the wellbore with the hydraulic rotation of the string, Kelly, and rotary table. He stops the rotation, unscrews and lifts the previous joint out of the hole with pipe **elevators. Slips** hold the **drill pipe** securely in the rotary table. The next joint is ready in the **mouse-hole.** The Kelly is disengaged from the drill string, swung over the mousehole, screwed into the new joint, and then raised over the slip-supported drill string where it is screwed into the entire string. The slips are removed and the process repeats itself.

At the working end of the wellbore lies the **bottom hole assembly.** This holds **drill collars** that add weight to the end of the string for directional control. Heavier drill pipe may have been added right above this connection for further stability and to prevent breaks. Along the string, **subs** are segmented in. **Stabilizers** keep the string centered in the wellbore; **hole openers** enlarge the bore itself, while **crossover subs** connect two different pipe thread types. The real work is done with the **drill bit.** There are more than 1,200 different types, shapes, sizes, and components. They can be fixed or rotary, directional or straight line. Their faces can be diamond, manmade diamond, polycrystalline diamond (PDC), or milled. They can last a few hours or hundreds of hours.

When they wear out, the driller can feel it—and the drill rate declines. He has to "make a trip," or pull the entire drill string out of the hole, change the bit, reassemble the entire string, and drop it back downhole. Tripping out the drillpipe entails unscrewing a treble joint, stacking each pipe length as it is withdrawn from the wellbore, attaching the new bit, and then tripping in. This can take hours, depending on the depth of the well.

In support of the drill string, the **circulating system** flows the **drilling mud** down, through, and out of the wellbore. The mud cleans the bit and brings the cuttings to the surface, keeping the hole open, and lubricates and cools the bit and its teeth. The mud also maintains the pressure differential necessary to keep the hydrocarbons down while the operation progresses. By caking the inner walls, the mud helps to stabilize the well sides and prevent blowouts.

The mud flows from the **mud tanks** right off the derrick pad, through the **agitator** and the **mud hogs**, down the wellbore, and through **jets** in the drilling bit called **watercourses**. It runs back up the **return line** to the **shaker**, where the coarser pieces of rock are separated. The shaker may have dual screens to filter finer cuttings as well as the larger rock chunks. The mud may flow through a third filter—the **desanders**—then back to the mud tanks for reuse. The tanks are baffled into compartments: shaker tank, reserve tank, and suction tank. A **mud-gas separator** will remove any gaseous hydrocarbon dissolved in the mud. A **pit** has been dug off to the far side away from the rig and the shakers to receive the cuttings. If you search through this debris, you may find shell remnants from 100 million years ago or more.

The drilling mud itself is designed to the specific well. It can be water-based, oil-based, a mixture of both, or synthetic. The water is typically highly saline—far saltier than the Jordan River. Oil-based mud is the best for lubricating the bit but can be difficult to dispose of properly. Synthetics and water-based are far easier to dispose of, but offer less lubrication. Bentonite is the typical water-based additive. It is volcanic clay with good suspension and viscosity. A wide variety of other ingredients may be mixed into the mud, as each well demands: bactericides, emulsifiers, foaming agents, thinners, and older cuttings are used in addition to the preferred mix of galena and barite. All of these "recipe portions" are stored in the mud house and added through the **hopper** by the **mud man**. The water weight can be increased from the normal 8.3 pounds per gallon to more than 20 pounds. The weight aids the drill bit by adding pressure at the bottom of the well. This weight is constantly monitored and adjusted.

Suffice it to say, rig operations are dangerous. The repetitiveness and monotony of the work can lead to injury as attention to detail slackens. These are very large pieces of heavy steel pipe rotating at significant speeds. The equipment is massive, with loud and dangerous engines. The best drillers can break out and make up pipe in less than a minute. Seeing this work in action, you can't help but be impressed by their "quiet" skill. If a roughneck is slow, or drops a tool or makes an error, he has to survive the derision of his mates. There is no room for error here; when you screw up, everyone knows, and you will have to take the heat. Nicknames are given, and they stick. You do not want to be known as *shorty, loose hands, slippery dick* or *tongless*. If the driller neglects to turn off the drilling mud pump, everyone gets a very warm bath in high-pressure water loaded with rock, soil, and mud. If a joint slips from the Kelly, watch out. If the monkey board man drops something, or falls or slips, there will be injuries.

Given these dangerous conditions, the continued decline of injuries reported to the U.S. Department of Labor is nothing short of amazing. This is a testament to the drilling crew's professionalism. Communication is key to the success of a drilling operation. The roughnecks working the tongs and slips must coordinate their activity. The noise makes speech impossible much of the time. You have to know what is going to happen each step of the process. You have to be aware of everything around you—above, below, and behind, as well as in your hands or at your feet. You have to be strong, agile, accurate, and smart. There is no room for "six sigma" error—no mistakes, which could cost an arm, an eye, a life. Each set must be perfect—or you will hear about your error for days.

While the talk can be rough, it is also specific and courteous. Every man knows his task, and does it with silent perfection. The admonishments may be severe. The "atta boys" are left unsaid. You are expected to do your job right.

The **toolpusher** is the site manager. He runs the operation, writes up the daily log, manages the workers and workflow, and orders equipment and trucks. The **driller** operates the equipment from the console on the derrick floor, which is usually three stories high. He orders the crew at their tasks and runs the console: the rotary table, mud pumps, chain drives. He monitors pressure gauges, tachs, pump stroke, weight indicator, torque indicator, and rate of penetration. The **derrickman** is second in command. Two to four **roughnecks,** or **roustabouts,** handle the equipment and maintain all of it. Rigs run in three eight-hour shifts with a full crew. Depending on the size of the operation, these crews can number from a dozen to a few hundred men. Most OFS operators move with their rigs. Punching a hole can take three days to three weeks, according to its complexity.

To the casual observer, the well hole may seem to drop perpendicular from the derrick. It actually falls in a **long helix**, which very slowly widens as the hole deepens. The deviation from the perpendicular may be as much as five percent. The bit walks to the right as it pierces the earth, corkscrewing down. Most operator contracts stipulate the well deviation. This corkscrewing is planned for and taken advantage of by the driller. As a slant is anticipated, the wellbore is oriented in the direction of the slant with this cyclical twist.

As the well is drilled, a **casing** is set. The casing is just that: a tubular wall set down into the hole. This stabilizes the bore as it penetrates the earth. It also initially protects the local aquifer—as deep as 800 feet—from wellbore loss of cuttings, liquids, or gases. Below this level, the casing continues to act as a shield and rampart against the pressure of the surrounding rock. These casings are cemented into the bore through **string joints** that vary between 4½ inches and 36 inches in diameter, depending on where in the bore, what type of bore, and the rock formation it is holding against. The casing is laid in a manner similar to the drilling string.

The string is broken out. The casing crew runs down a **wall scratcher** to remove **filter cake** with its protruding wires. The casing joints are run into the well, down to just above the current wellbore bottom. The **slurry**, typically Portland cement, is prepared according to the well type, its depth and pressure, and setting time for the cement. The **cementing head** is attached to the Kelly to receive the slurry from the cement pumps, and a **wiper plug** is run downhole. This bottom plug separates the drilling mud from the cement slurry. A **float collar** stops it just above the

guide shoe, which rests at the well bottom. This "floats" the entire casing string in the drilling mud while allowing the slurry to fill the gap between the casing string and the naked wellbore. The cement is pumped down the wellbore and flows up between the casing and the earthen wall, the outside of the well. The second plug—the **top plug**—floats down with the slurry mix. When all of the slurry has been displaced to the outside of the well, the pumping is halted and the slurry allowed to set up. The plugs, shoe, and remaining bottom hole cement are all drilled out.

Casing is done in stages. As the well deepens, it is cased, then drilled deeper and cased again. The casing slowly declines in diameter as the well drills down. In addition to the entire well casing, three or more other casings may be drilled. The **initial casing**—the conductor pipe—has already been discussed. It runs the entire depth of the water table, or aquifer. The **surface casing** runs down the next thousand feet of wellbore. In addition to strengthening the hole, it protects the deepest aquifers. **Intermediate casing** can protect against difficult zones, such as soft shale and high pressure (the Morrow Sands in New Mexico, for example). **Production casing** runs the entire length of the drill string. The bottom can be finished in a variety of ways, but **set-through completion** is the most common for the shale gas and oil frac rigs we are concerned with here. This is the final casing slurry feed.

If you examine a well hole from the vertical, at the top you will see as many as five or six casings, the annulus, the widest being the conductor pipe at 36 inches. Farther down you may see two to four casing circles, depending on strength requirements and well depth. The drillers are creating a solid circular wall as deep as 16,000 feet. Think of these as the concentric rings of a medieval castle. The bore itself is the castle keep, the stronghold protected from without by their defensive barriers. The casing rings and their proper deployment are the major defense against seepage into or out of the wellbore. While the presence of a single casing may allow for the escape of gas at a rate of two to five percent, in the Marcellus, among other places, this drops to virtually zero with additional layers of concrete and casings. These casing layers are now mandated to ensure the methanes release approaches zero. It is the most lethal of greenhouse gases. While its presence in the atmosphere dissipates in less than two week, it is 50 times more dangerous than carbon dioxide.

Horizontal Drilling

The preceding section described in detail drilling a vertical shaft; however, if we are going to frac, we often do so today by going horizontal. It is difficult to imagine these cemented-in castled walls deep in the earth as retaining flexibility, yet they do. The driller knows the depth at which he must begin to "turn the bit," the **build angle** he can do so at, and the amount of carry downward—the **drop angle**—he can expect before he is finally horizontal. He wants to change his angle of attack because he is looking for the maximum **seam penetration** possible for the wellbore. Imagine you have a flexible straw, and you want to drink only the liquid in a glass full of ice. If you push the flex end down into the liquid, it begins to turn horizontal as it contacts the bottom of the glass. Now you can drink what is below the ice. The driller is doing the same—substituting rock for ice. He has to **"kick off"** the bit gradually as he approaches his horizontal target.

Kick off for a wellbore has been with a **whipstock** in the past. This long angled-steel wedge alters the direction of the drill string. It is cemented in place before a **mill** is used instead of the bit to break through the well casing and change the direction of the wellbore. Once the angle has been defined, the driller can change direction as the seismic indicators direct him. It takes time and patience. Today, the downhole assembly includes a **bent sub**, a steerable **turbine mud motor,** and a **diamond bit.** The bent sub, a flexible steel joint, angles the turbine to the correct drop angle, and the turbine is driven from above in real time as the driller follows a predetermined geosteering section of the deposit as a 3-D map. Today, **coiled tubing** replaces 30-foot lengths of steel pipe casing, adding flexibility, continuity, and lateral strength. The **build angle** is precise.

Once the horizontal section is reached, the driller runs the casing, or **coiled tubing,** out the length drilled by the mud motor. The greatest distance for such a horizontal reach is more than eight miles, and the driller can "unlock a suitcase" at the end of the drill string. The casing continues to be placed at intervals, or the coiled tubing is laid, to secure the well against the now-enormous pressures at depths of 4,000 to 16,000 feet . This horizontal well can be completed in less than three weeks, running three shifts.

The driller sits in a comfortable room, driving the bit much like a young soldier might drive a drone over the hills of Pakistan, seeking a kill zone. The driller seeks a thin length of hydrocarbon several miles down, and away from him. He finds it every time. He leads the coiled tubing to its precise spot. The precision of tunnel drillers, such as those who constructed the Euro tunnel, is taken much further in these deep shale formations. Drilling techniques today can turn in several directions to envelope the reservoir in frac seams.

These wells tend to have extraordinary production rates. While the initial flow quickly tapers off in a few months, it maintains a level of 20 to 10 percent of the original flow, often for years. When you have initial daily gas flows of one million cubic feet per day (Mcf/d), a resultant flow of 200,000 to 100,000 per day is very acceptable.

Fracing

We have done the geology, we have found the shale source, reconnoitered its depth and breadth, acquired the land, cleared for the pad, brought in the rig or constructed the derrick, punched and cased the wellbore to the production zone, and are ready to prep and perf the well. This is the beginning of the frac. The well, either a new one or a workover, stands ready. The toolpusher has called in the frac rigs. Massive trucks wind their way up to the pad site: wireline services, water trucks, sanders, proppant trucks, diesel fuel carriers, and disposal trucks.

The following is an abbreviated discussion on a topic that is both extraordinarily complex and evolving, in real time. Knowledge in the industry and in the field grows geometrically, every month. Recall how hydraulic fracturing and horizontal drilling are still in their infancy. Mitchell was doing his work in the late 1990s, but application of his acquired knowledge only began in earnest less than a decade ago, in 2003. If you want to find a career in a cutting-edge applied technology field, you would have to look long and hard to do better than the unconventional oil and gas fields of America.

Perf is simply the preparation of the well for the **fracing stages.** The perforation of wellbore casing varies between well sites, companies, fields, and nations; in fact, there are very few commonalities from one to another. Some firms take the "one-size-fits-all" approach; others vary the practice even within a well. As we have seen earlier, the manner in which a well is fraced falls into three categories. The types and varieties of **proppants** increase every year. The size and number of sequential fracs can vary with the complexity and depth of the well. What follows is a simplistic description of the most dynamic aspect of the oil and gas industry.

Plug and perf, the dual process of perforating the wellbore casing and plugging the resultant hydrocarbon released pressure until the entire wellbore is ready for the fluid flow, are both done by the **wireline service** crew. In brief, a gun barrel is dropped down the well hole, and shots are fired through the casing and into the rock formation to break open seams. Jake's team showed us how it is done. These seams will next be **fraced**—saltwater and sand will be forced into them—to extend their reach deeper into the deposit. These shots are fired not from the front of the gun barrel, but from the barrel wall itself. The wire service crew places the barrel, pulls the trigger, and removes it. The frac team prepares the hole and fracs it. These steps are repeated multiple times to open the reservoir for extraction.

The details can be daunting. Shale deposits vary within each well and across a production zone. The perforations created by the shots may be closed by pressure, have excess debris that clogs, or may not fully stimulate the potential reservoir. Acetic fluids are often pumped into the well to clean debris from around the **well shoe,** or foot of the well, before the frac stage begins.

The wireline servicing industry constantly evolves it solutions as it encounters new problems. Approaches to perfs today can be defined as **cemented, open hole,** or **multi-stage;** each has pros and cons. Guys never tire of talking about explosive charge size, how the charge is shaped ("she's got a pair of hips…"), and the source of explosive. These guys talk like the explosives crew on a movie set. Their work is far more dangerous.

Reactive metals in today's perf charges cause violent mini-explosions that drive the perf tunnels deeper into the earth. Shot sizing, staging, and density varies according to rock demand. Five-foot lateral reaches by these charges are common today. That is 2 ½ times the reach of just two years ago. This opens the volumetric size of the play considerably. Imagine a cucumber. Now increase its size to a watermelon. That is the shape of the frac zone.

Think about this for a moment. The gun is two miles below your feet, and as much as 40,000 feet to your left or right at the end of 100 tons or more of steel tubing. The heat and pressure at these depths is enormous. The shots are fired simultaneously across dozens of openings in the barrel. The explosive tunnels penetrate 60 inches or more into the solid rock. The precision of this process is critical to its success.

The gun barrel is removed, a cement plug is filled in to cap the section, and the next 50- to 150-foot section is ready to be perfed. This process repeats itself until the entire production area has been perfed and each section plugged. The gun is retrieved, a special bit mills out the plugs, and the cuttings are washed up and out. Plug and perf can be done across dozens of stages in a wellbore. This process usually occurs in stages over a few days. If it happens during one continuous operation, it is known as a **multi-stage frac**. These are the cutting edge tools, the sc-fi, of the frac industry. These advanced perf logs can run dozens of stages simultaneously.

There is no industry consensus as to which to use: plug and perf, open hole, multi-stage, or these newest solutions, **perforating sleeve** technology. These sleeves do the frac work simultaneously across as many as 40 intervals, or **stages**, a waltz of small but incredibly powerful explosions, perforations, and penetrations. Perforating sleeve technology is the latest attempt by the biggest, most capitalized OSI firms to develop a reservoir as quickly as possible. It does in a single pumping operation what had previously taken days to accomplish. In one trip it can replace staged plug-and-perf operations, as well as the subsequent milling of the cement plugs. The process can also use far less energy and water than plug and perf, resulting in less impact on water tables, less need for water truck importation, fewer truck miles, less air pollution and far fewer chemicals.

Though their work is repetitive, every drilling crew can count on problems arising on a daily basis. **Tortuosity** is just one of many. The perfs penetrate solid rock by creating cracks, or fractures. These should be as direct and open as possible for the hydrocarbons to flow upbore. Tortuosity—the nature of the rock to compress, twist and reduce volume between pore spaces—can virtually eliminate perf effectiveness; among other things, it can torque sleeves.[75,76]

Pressure should be working to the benefit of the crew, not against it. There's no single, industry-wide solution to this problem. Costs, worker skills and experience, available technology, downhole geology, and capital availability and requirements—each of these choices directs the operations. Together, they present the applied technology challenge at the leading edge of fracing.

The Estimated Ultimate Recovery (EUR) from the production zone defines the final test for success. The EUR is itself a direct result of oil and gas prices further downstream. The higher the price for the hydrocarbon product, the more profitable is the extraction process. Thus, more capital is available to enhance further the process and its production. Oil at $54 per barrel results in a lower EUR than oil at $150 per barrel.

The frac is done in intervals. The frac is, essentially, a forcing of briny water under very high pressure down the wellbore into the fractures in the production zone. The water is laden with the sand. When the water is pumped out, the sand remains, holding open the newly forced seams in the deep rock. Hydraulic fracturing is just that: using water to fracture rock.

When it comes time to frac, dozens of large trucks have been stationed in neat rows at the well site. Pumping trucks deliver frac fluids from tanks to the **wellhead manifold**—the top of the capped well. A million gallons of brine may contain 3 million pounds of proppant. The pumps—single through quads—may force 16,000 pounds or more of pressure down the borehole. The manifold itself is protected against both the pressure and the corrosion of the liquids.

Chemical additives are laced into the brine and sand admixture. These additives are often a trade secret. Figure 20 shows the constituency for one frac job.[77]

Understanding Fracturing Fluid

The fluid from the hydraulic fracturing process is nearly **99.5% WATER & SAND.**

9.5% Sand

0.5% Chemical Additives

90% Water

Typical Additives Used in Fracturing Fluid and COMMON HOUSEHOLD ITEMS

SODIUM CHLORIDE
used in table salt

ETHYLENE GLYCOL
used in household cleaners

BORATE SALTS
used in cosmetics

SODIUM/POTASSIUM CARBONATE
used in detergent

GUAR GUM
used in ice cream

ISOPROPANOL
used in deodorant

To create productive natural gas wells, companies force fluid thousands of feet below the surface at high pressure to crack shale rock and release trapped natural gas. This extraction technique is called hydraulic fracturing. The fluid used in the process is made up almost entirely of water and sand. However, it also includes a very small percentage of chemical additives that help make the process work.

All of these elements are present in most households[78], but in large concentrations, they can be dangerous or even lethal. Their restriction from local water supplies, grounds, and deep aquifers is absolutely critical to the success of each and every well drilled. Their accidental release requires immediate notification and remediation. This is serious business for all concerned: operators, regulators, landowners, and NGOs. While the amounts used are small, their presence, or accumulation, is a concern. Regulators, the industry and the local community pay close attention to the details of the mix, when and where it goes, how it is recovered, and, if it spills or escapes from the wellbore, remediation efforts – both immediate and ongoing. This is the stuff of environmental and legal nightmares. Prevention and wise use carries a hundred times the value of clean up.

The frac charges are set. They follow in their exact firing sequence. The fractures open, and slick water flows in to force them open ever more deeply. The water withdraws in a tidal bore, leaving small particles of sand to hold open the fissures. These capillaries soon run flush with the gas, or oil, toward the surface, toward the pipeline, toward the fractionators, processors, and storage tanks. At your utility company, it fills pipelines that drop the precious gas into boilers—the boilers that spin the turbines that produce the electricity that moves through its own pipeline to your home to heat you, feed you, cool you, and bathe you.

PART TWO:

Opportunities and Challenges

Fraced wells give us extractives for which we have many uses and opportunities. Fracing also presents challenges, which both the industry and opponents are quick to point out. With a nod to both proponents and opponents, but an endorsement for neither, let's examine both the opportunities and the challenges associated with fracing.

OPPORTUNITIES AT THE LOCAL LEVEL

Natural gas is better distributed than any other fuel in the United States. Down every street and up every alley there's a pipeline.

—T. Boone Pickens, American oil tycoon

Once extracted from the wellbore, where do hydrocarbons go? How do they get there safely? What happens to them upon arrival? These are our next areas of investigation—the distribution, storage, treatment, transshipment, and ultimate use of the oil, gas, and NGLs.

For simplicity's sake, we shall focus on natural gas, or methane, with the understanding that crude oil distribution is similar in some respects but has its own complex world of pipeline, fractionating, storage, and user needs.[79]

Pipelines carry gas and oil from the wellhead to the utility to the world. There are literally tens of thousands of miles of pipelines beneath our country's surface. Figure 21 shows the major lines crisscrossing the U.S. as of May 2008:

Natural Gas Pipeline Map

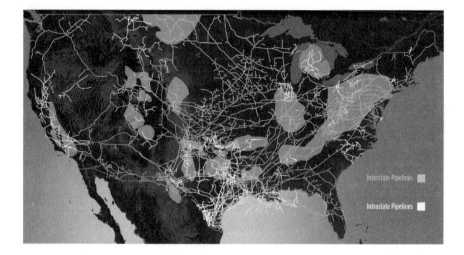

These pipelines carry everything from crude oil to aviation fuel and are generally classified for gas or liquids. The pipelines are further classified as either gathering lines or main lines. The wellhead stream itself can be gas, liquid, or a mixture. If it is a mixture, **field separators** segregate the oil and gas, and these are distributed independently. **Compressors** maintain pipeline pressure along the route to ensure a constant volume and velocity. The gas flows from lower pressure to higher along the pipeline as the compressors inject new gas into existing flows. The gas is **fungible**—interchangeable—so it can be intermixed and exchanged across a fairly wide spectrum of gas sources with different Btus. If the wellhead pressure is high enough, the gas is simply added to the flow.

If the gas emerging from the wellhead has no hydrogen sulphide or water, it is called **sweet gas**. When HS2 is present, you have **sour gas**, an extremely dangerous substance that "kills in two steps." Sour gas, therefore, must be stripped of its poison. First, the water is removed since it will corrode the pipelines and quickly reduce flow characteristics. The waterless sweet gas exits the separators, the sulphur dioxide is stripped out and the methane is metered into the gathering line. Next, it is sent to the gathering and processing plants. Here, the heavier **NGLs— ethane, propane, and butane**—are removed. The **cryogenics and silica gel plants** strip out the heavier hydrocarbons from the methane, which then re-enters the feeder pipeline for bulk distribution. The NGLs are themselves stored or transshipped downstream.

The processed gas is sent to the local distribution companies (LDCs) for final distribution to residential, industrial, or commercial users. The gas going to utility power generation is delivered through larger-diameter pipeline. Industrial or commercial bulk users receive the gas either as **flow-controlled** or **pressure-controlled**. Flow control limits use to a maximum withdrawal rate. Pressure control allows the end user to withdraw as needed, when needed. The flows are priced accordingly.

Distribution and Storage

The gas or oil can be sent to final users or stored for future use. These storage facilities may be above ground or underground. Tank farms are collection-area storage facilities—many tanks banded together on a plot of land. Underground storage for gas is often in depleted reservoirs, salt domes, or caverns. Much of the gas placed underground is stranded—as much as 50 percent of the gas in an old reservoir or a salt dome is simply **cushion gas**, only there to maintain sufficient pressure. If the pressure were to drop significantly, the well walls may collapse and block further flow.

As demand varies with a certain degree of seasonal predictability, operators can fill and deplete with some normality. Down time for fractionators or line maintenance can be managed with seasonality in mind. Summer storage tends to fill up in anticipation of slackening demand for heating, which then resumes with increased demand for air conditioning. Weather concerns, such as a heat wave or a hurricane, have a direct and immediate effect on both demand and storage.

Storage capacity is used to manage demand swings. The distance from a storage source also impacts delivery. It takes four days for a methane molecule to travel from the Gulf Coast to New York City, and another day to Boston. Storage reports often show significant increases and decreases with the seasons. If a political issue arises, demand pulses can overwhelm delivery systems in the short term. Given the massive amount of gas currently available, and the continual increase in the flow from wellheads, these demand alterations are having a far smaller impact on the demand cycle than in previous years.

Central control rooms regulate access to the lines to meet demand fluctuations by "packing" or "unpacking" the line, or filling it with more product. Demand is a function of time, weather, season, economic conditions, and pricing. The control rooms monitor hundreds of delivery and receipt stations. Operators want as high a "line pressure" as is safely practicable to reduce line friction and to ensure efficient delivery. Plant shutdowns, severe weather, and holiday demands are examples of events the operators must be prepared handle.

Breaks, line ruptures, and equipment failures can affect much of a delivery line. According to the U.S. Department of Transportation's Pipeline and Hazardous Materials Safety Administration (PHMSA),[80] the most common reasons for delivery line failure are breaks from excavators and agricultural activities; natural hazards; corrosion; and mechanical problems.

Preventing these is the duty of each pipeline operator, and is a constant challenge as two of the four reasons for failure are entirely outside of their control. The latter two they can control through constant inspection. Internal inspection devices called **pigs** are run downline to proactively detect potential problems. Lines can also be swept and cleaned of deposits. Pigs are actually metal or plastic spheres launched and received through special **pig lines**. The receiving end is of a larger diameter than the surrounding line to allow the pig to slow down and drop into the receiver's traps. Visual inspection can be performed on aboveground line, while magnetic, ultrasound, hydrostatic and electrical tools, and surveys are used for most lines.

The two recent breaks in Enbridge's major oil pipeline in Wisconsin have local communities, the State of Wisconsin, and the U.S. Department of Transportation (DOT) furious. DOT has yet to allow the pipeline to reopen because of the "Keystone Kops" attitude of corporate management, according to Ray LaHood, DOT Secretary. The sudden increase in demand for pipeline may cause similar challenges for the industry as these demands are met with thousands of miles of new pipelines.

More lines, working the right fluids, going to the right end users, coming from the correct major lines, are all required—and notable by their absence. The growing pains of the industry must be matched by the industry's skills and experience. Communication with the public and regulators must be timely, honest, and forthcoming. The lessons from Wisconsin should be applied to all oil and gas firms seeking to expand. This expansion is not simply about pipelines and profits. It must be an expansion of public awareness, acceptance, and understanding.

The technically recoverable shale gas in the U.S. is now estimated at 1,700+ Tcf, or 50 years of energy production's need.

Get it right each time a new line is installed, and you win the public over. One mistake hurts the firm and the industry. Repeated problems can be devastating. Regulatory delays only compound the difficulty for installers and controllers. On this issue, the current administration, in the form of DOT's Mr. LaHood, seems to be trying to make the system work, and Canada's **Enbridge Energy** is the drag on corrective measures.

The pipe must be installed. This involves permitting, landmen, leasehold arrangements, rights of way, and assembling the pipeline **spreads**, or installation crews and their equipment. Topsoil is stripped and stored for replacement. Bridges, tunnels, and easements are constructed to avoid public traffic. Stringing the pipe can occur across desert, farmland, riverine, mountain, fields, and urban landscapes. Each brings its own challenges.

Capital is available, certainly. The increase in capital invested in midstream ventures was 200 percent for 2012, as the need for build-out is met with investment from private and public sources. Global capital inflows through 2030 should exceed $10 trillion in new-generation projects.[81] This is in addition to the $7 trillion in upstream operations capital.

The pipe itself is manufactured by pushing a hot steel cylinder over a solid rod, or **mandrel,** that hollows out the cylinder, forming a seamless pipe with a beveled end. This allows the weld bead to penetrate the pipe joint wall. The pipe is welded to its subsequent mate by one of two welding techniques, either **submerged arc** or **gas arc**. To complete the weld, the **root bead** is followed by the **hot pass, filler bead**, and then a series of **cap beads**. The weld is inspected using radiographic or ultrasound observation. The steel pipe is coated for insulation, corrosion, and electrical resistance. Joints may be taped or shrink-wrapped. Changes in direction of the pipe in size and in function demand fittings, flanges, valves, and actuators. These are added onsite to the line where required.

Trenching is done with a backhoe or trenching machine. Old-fashioned hand trenching occurs around obstructions and natural habitat. The pipeline is lowered in once the trench is prepared, and the welds are inspected. The backfill completes the process. Bends, crossings, and tie-ins of other lines allow them to bypass rivers, roads, and structures.

The complexity of pipeline design, maintenance, and flow management requires the highest degree of skill and experience. Though unknown and invisible to you, these men are part of your daily life by allowing the flow of gas and oil to your home light switch.

Safety is a culture that most firms design into their operations. Whether they are welders, installers, operations, control-room technicians, or backhoe operators, safety is a priority. As a result, there are fewer reported accidents in the pipeline industry than in any other form of transportation.[82]

The public, however, does not distinguish between pipeline firms, or even between energy companies. An incident makes news, and sometimes the truth gets buried in all the noise, especially with today's incessant online chatter. To help avoid a public relations nightmare, all members of the pipeline industry and its field support must understand and be able to articulate industry safety and performance standards, including monitoring and training activity, reaction capabilities and response times to incidents. Involving the local community in decision making also is important. The pipeline goes through once, then disappears beneath the soil. Involving the community during installation ensures communication, understanding and approval.

As the exploitation of unconventional gas and oil reservoirs continues its inexorable progress, **health, safety, and environmental issues—HSE** in industry speak—remain at the forefront of industry activity. They should also be at the forefront of communication within and among firms and industries, as well as all stakeholders including owners, leaseholders, citizens, municipalities, states, suppliers, competitors, employees, the media, and NGOs (non-profits). This involvement during each stage of exploitation and enrichment empowers local communities and other stakeholders to voice concerns, participate in discussions, and rightfully hold the ultimate decision makers accountable.

Natural Gas Use

Pipelines meet at hubs. Major U.S. hubs include the Henry Hub, Cushing Hub, and Opal Hub. Hubs are gathering arenas for storage and transshipment, whether marine or land. Storage capacity is tremendous. Hubs are the heart of the hydrocarbon market; at hubs, pricing is established and adjusted every minute of every day.

The Federal Energy Regulatory Commission (FERC) controls interstate commerce. It also approves **tariffs** for transit, which are the capital lifeline of the pipeline industry. The pipe is like a toll road along which the molecules pass, for a small fee. Most midstream firms today are publicly traded MLPs. They earn their income from tariffs. They pay their overhead, and reward their shareholders with dividends earned from this tariff. They develop new capital in the equity markets by offering new shares, occasionally bonds. These firms tend to be strong dividend payers because of their significant free cash flow.

As you may have noticed from the map at the beginning of this section, dozens of pipelines are crisscrossing the U.S. The natural barriers of mountains and population centers break these flows. Barges move much energy source across the Midwest. The Rockies allow only a few pipelines to cross their lofty shoulders, which means the western states are heavily dependent on California's abundant production fields—and the political restrictions imposed by that state. There are literally two energy markets in the US: the western littoral and the rest of the US. Only five pipelines cross the Continental Divide. Virtually all energy used on the West Coast comes either from California or Alaska.

Natural Gas Liquids—NGLs

Refined products are known as **natural gas liquids**. They are removed from the methane gas by the **strippers**, then typically stored and shipped in batches. Sometimes there is a unitary system for an area, such as the Yellowstone Pipeline. This receives product from each of the local refineries and ships it to terminals for ultimate distribution to commercial, ranching, and residential end users. Most often, the gathering and distribution lines are multi-staged and loci-centered. The gas is shipped to staging tanks for storage and processing before being sent on to industrial users. These NGLs are three simple hydrogen/carbon molecules, whose chemical names are **ethane** (nine to 13 percent in natural gas); **propane** (three to four percent); and **butane** (one percent).

Ethane is used as a **petrochemical feedstock** for ethylene, the core of much of the plastics industry. All of the NGLs are separated out from the methane gas through the **cryogenic separator.** Here, the temperature of the gas is dropped to -100 C. The NGLs liquefy and are **fractionated**, or distilled, from one another, and then stored. The ethane can be used as a refrigerant or further distilled, by steam cracking, into acetic acid (recall its use for cleaning the perf wellbores) and vinyl chloride. This highly toxic substance is further catalyzed into PVC, a type of plastic. One of its obvious applications is the PVC pipeline in your garden and home. It is also feedstock for plastics production. PVC is a completely stable compound whose use has no hazardous consequence. Ethane and ethylene have no known acute or toxicological risk other than severe frostbite if you come in contact with it. Ethylene is a major component of oxy-acetylene welding equipment and is also used for ripening food. In the Marcellus, **Keystone Midstream** is mixing the high-quality ethane with a smaller portion of methane to produce the fuel for all of its engines onsite—an ideal recycling effort.

Propane is used by more than 10 percent of U.S. households for heating and cooking. Your barbeque and your clothes dryer may be propane-powered. It is used in agriculture for crop drying, weed control, and vehicle fuel. It is used in off-grid homes as a source of heating, cooling, weed control, and cooking. On the lighter side, it is used in hot air balloons, paintball guns, and as the explosive used for Hollywood's special effects. Industrial and manufacturing plants are the primary users, followed by home and transportation use.

In transportation, as liquefied petroleum gas (**LPG**), it ranks third in the U.S. for fuel, after gasoline and diesel, operating more than 250,000 vehicles and 300,000 forklift trucks. In metropolitan areas across the nation, 100,000 buses, taxis, and delivery trucks run on propane fuel. It is a cleaner-burning fuel with lower GGEs than gasoline or diesel. It has a higher octane (104 to 112) compared to gas, so its use extends engine life, often doubling it. Cold-start problems, a quirk for the northern driver, are a nonissue with LPG.[83]

Most propane is stored in massive salt domes or pressurized steel tanks. It is transported to end-use facilities by tank car, truck, and barge, but primarily by pipeline. Its storage is always under pressure, which greatly increases its usefulness and productivity. Safe storage and use continues to be an issue, and the propane industry takes pride in its ongoing training for safety and safe handling.[84]Much of it is now exported through the Houston Shipping Channel and the complex of terminals there. Your author has watched an 800-foot tanker being filled with propane in less than a day. It was outbound for South America. The lines coming in to the field and transferring to the ship were a warren of pipes, with storage tanks encircling the terminal. More pipelines and storage tanks were being added every day. The demand was endless, the construction never-ending.

Derived from the same cryogenic process as ethane and propane, butane is a more complex hydrocarbon compound. Shipped and stored similarly to ethane and propane, it is primary used for heating and transportation. When mixed with propane, it becomes LPG. It is the primary propellant in aerosol spray cans, replacing halomethanes, banned for their ozone depletion properties. Butane is also used for heating and cooling, and powers camping stoves, small torches, and portable space heaters. Butane was the fuel used to send into orbit a small satellite developed by the British in 2000. For all its uses, butane raises toxicological concerns—it's the leading cause of "solvent sniffing" illness and death. The initial sense of euphoria resulting from its misuse as an inhalant is extremely dangerous and narcotic.

WORLD OF ALTERNATIVE ENERGY: OIL AND GAS AND NUKES—OH MY!

There is no alternative way, so far discovered, of improving the lot
of the ordinary people that can hold a candle to the productive
activities that are unleashed by a free enterprise system.

—Milton Friedman

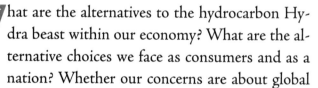

hat are the alternatives to the hydrocarbon Hydra beast within our economy? What are the alternative choices we face as consumers and as a nation? Whether our concerns are about global change or local consumption, choices enable us to make informed decisions.

Nuclear energy has been considered for many decades an ideal choice for energy production. President Eisenhower felt that once we turned to nuclear power for our energy there would be no need for meters in our homes – the electricity would be essentially free. It is clean, safe, and cost-free once the massive costs of design and installation are achieved—that is, if you factor out the opportunity cost for maintenance, storage and environmental impact.

These expenses add to the capital expenditure for any firm looking to develop nuclear energy. The French know how to build and maintain nuclear energy plants, store the end product, and price the electricity. Of French power generation, 80 percent results from the controlled splitting of the uranium atom. The Russians, Japanese, and Germans know these same applied-technological principles. Along with the U.S., they are leaders in this energy technology. Yet, the political levers exert more power than that generated by mere atoms. Germany and Japan have retreated from nuclear power generation after decades of fiscal and political capital development. In the U.S., no new new nuclear power plants have been built since the 1970s. The Russians do what they choose. The story behind the political intervention into nuclear power is beyond the scope of this book (read Daniel Yergin's *The Source* for the best account). We have simply to acknowledge that the political will to support such power sourcing has evaporated.

Alternative fuel sourcing for power generation, however, has the current political backing to succeed. It lacks the capital power to stand on its own. Forty years or more of development have yet to bring the cost of energy production down to the competitive levels of other energy sources. Economic competition is fierce. As long as we live in a capitalist society, the economics of power generation should ultimately hold sway.

There will be no further nuclear power generation, at least in America, until its costs are in line with risk expectations. The new political-risk equation removes this power source from any conversation. So be it. Whether that is good or bad is ultimately a **life-cycle systems analysis** decision: all aspects of the chain of power generation must be evaluated and priced before it can move forward once again.

Life-cycle systems analysis involves the throughput of all aspects of costs and benefits for a given energy source, and its production and use. How this equation is stated depends on the context. At its simplest, it includes the costs of exploration, production, distribution, refining, and transshipment, as well as the current and long-term social and environmental impacts.

This equation is easier stated that explained. Each of these units, from exploration to impact, within the equation is cut from whole cloth. Each represents an entire complexity of other equations. Defining social and environmental impact is still a work in progress. We have only recently begun to ask questions, so there are no real answers and, certainly, there is no consensus. Mathematicians do not suffer fools for long. These issues must be defined before they can be applied. NGO players are applying serious efforts to answer these questions. These groups wish to enforce a more global understanding of the economic, social, and environmental impacts of "simple" economic decisions. To their credit, some are working hand-in-hand with the oil and gas industry to affect these cost analyses. To the credit of the involved firms, recognition is slowly being given to these real and potential long-term impacts. Jointly, these workings will produce a new economics of scaling, appraisal, and evaluation. This new economics will be based in both capitalist and humanist values.[85] In an attempt to work cooperatively toward solutions to global and national energy policy conundrums, several groups have joined together to form the **The Bipartisan Policy Center**.[86]

In the meantime, energy policy dances on the blade of a knife, balancing between destruction and creation. The economics of life is a dance of *wu lei* masters, whether the dance is performed by single-cell microorganisms or by multinational, global, political, or corporate entities. The current **power sourcing and generation policy** is dominated by alternative energy sources, encompassing the sun, wind, soil, and water. Hundreds of billions of dollars, euros, yen, and yuan are being committed to research intended to break the **cycle of dependence on hydrocarbons.** A new world order of power generation has been imagined and is now being explored. Will these new Holy Grails be discovered? Will the search be in vain? Markets will decide – or government policy pressure will force the decision.

The success today is determined by the capital commitments from private capitalists and from government resources (read tax revenues). Many sources have commented on the ability of top-down research to work in the field of venture capital. The failure rates of venture capital are well-known, and the words "burn rate" have entered our language from the equity markets. These burn rates, coupled with these capital commitments, seek these new grails. Today, however, these grails are feeble and inefficient. They fail soon after leaving the capital incubators or must be constantly supported by further capital flow. None have yet succeeded in meeting or exceeding the cost potential and profit opportunities that a capitalist market demands.

The cost to produce a kilowatt of electricity in the U.S. in the fall of 2012:

Nuclear	<$.01/kWh
Coal	$.04/kWh
Natural gas	$.02/kWh
Wind	>$12/kWh
Solar	>$26/kWh

These figures exclude government support in the form of tax or regulatory relief. Nor do they factor in the cost of installation. They simply apply to the cost to produce.

Perhaps the only hope is a change in the rules. If capitalism is replaced with another metric, perhaps these new power sources can survive. If economics are defined more broadly, to encompass the long-term effects on society and the nation, perhaps these tools can work in this new world. Absent these systemic changes—the complete redesign of the economy, of our capitalist system—each of these alternative energy sources seems destined to fail. We would have to redesign the economic system away from a reward for trying, failing, trying, succeeding. The system would have to reward the energy process with the least impact across human culture. This system has a name: the precautionary principle. If you change the rules, if you change the system, then this may work. Certainly that appears to be the intent of the federal governments of many nations, including ours. If this is the course we intend to follow, perhaps these power-generation facilities will ultimately replace the hydrocarbon cycle. But the entire profit motive concept would have to be replaced with the "do no harm" system.

In the meantime, cost matters during the struggle for power pricing—locally, regionally, and globally. Tax support for alternative energy acts as sand within the lubrication apparatus of energy supplies nationwide. This friction reduces efficiencies. This friction destroys jobs and families and communities. As long as we subscribe to capitalist mechanics, this truism applies: Taxes destroy wealth. Tax credits in virtually any form are redistributive. They force an advantage where none lie.

To see how the sands of tax support muck things up, look what happened when Chinese solar panel firms received tax subsidies from the U.S. government to produce panels destined to be shipped to U.S. buyers. These Chinese firms have received tax subsidies from the U.S. government to produce a product that they then ship to U.S. buyers. Their prices are lower than local manufacturers. Active wholesalers purchase these and resell them to retail's end users. Lower cost drives consumers to buy these products. Now, the federal government is slapping a regressive tax on these buyers. Why? Their purchases disadvantage local suppliers. If you bought one of these Chinese panels, you now must pay a 'back tax' for so doing. This is your tax dollars at work.

Logic fails to explain the inconsistencies in this series of events. If you are the Chinese firm producing these panels, you have attempted to subscribe to the rules of the game, rules that allow you to sell in the U.S. market for solar panels. If you are a distributor of these panels to consumers in the U.S. market, you have attempted to follow the rules of importations and distribution. If you are a family that buys a set of these panels in an attempt to meet local regulatory guidelines, be energy-efficient, and cut your energy bill, you attempt to buy at a fair and reasonable price.

Each of you gets penalized for doing what you thought was right, according to the rules of the game you were playing. Each of you may be rightfully concerned or angry about the regressive cost imposed on your "wise" decision. Each of you may wish to demand a new playbook.

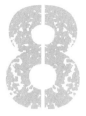

ELEMENTS OF PORTFOLIO DESIGN

Retirement at sixty-five is ridiculous.
When I was sixty-five I still had pimples.
— George Burns

This book was inspired by the author's inquiries into fracing from an investment perspective, so, to the extent it is relevant, I offer here a refresher course on my first book, **The 7% Solution**, which shows how you can design your retirement portfolio to meet your income needs. Here, we will examine specific content about portfolio design relative to the world of fracing. Has it opened up investment opportunities, especially for retirees?

The publicly traded firms that explore, produce, transship, process, store, and deliver for ultimate consumption the natural gas, oil, and NGLs that fuel our country and economy are often firms that happen to meet certain "retirement income" guidelines. These guidelines are:

- Strong free cash flow
- Low or no debt
- A history of rising dividend payments

- Sufficient trading volumes
- Clearly stated accounting

Many also have a component that is either quite attractive, or unattractive, to the individual investor: tax complexity—the notorious K-1.

As a result, any list of MLPs arrived at through the normal course of securities investigation represents a thoroughly oddball family. Whether you view these as misfits or geniuses is entirely your opinion. These are the facts.

Firms in the extended world of fracing are best represented at this website, *www.naptp.org*.

There, you will find the national organization to which most firms belong—most MLPs, or Master Limited Partnerships. This list is appended in the back of this book. It does change frequently, so please visit the website for up to date information.

The MLP is a creature created by Congress in 1981.[87] It was designed in answer to the desire for a legal entity which avoided the dual tax regime of the typical C Corporation (taxation of dividend at the corporate and individual levels), yet which paid dividends to its shareholders. The MLP was created—a space almost immediately occupied by a strange assortment of bedfellows and bedbugs. By the turn of the century, the sheets had been cleaned and the better folks moved in.

The MLP organizing structure today is the tool of choice for much of the energy industry. To quote a January 10, 2012 news release from the NAPTP:

A significant amount of the natural gas, crude oil, natural gas liquids, and refined products produced and consumed every day in the United States is transported by pipelines and stored in facilities owned by these MLPs. Approximately 25% of the propane delivered to rural businesses and homeowners are delivered by assets owned by MLPs. MLPs have also played a leading role in building the energy infrastructure needed to facilitate the production of new sources of U.S. natural gas and crude oil from shale and non-conventional formations.[88]

Today's MLPs occupy a respected segment of the investment world. The quadrant of the sky that they fill is bright with income. Its luminosity isn't clear through the viewfinder of pensions, charities, and trusts. It is best viewed by the individual income investor. This is a good feature—no investment black holes here. When you invest in an MLP, you can see the income they create as a check is sent to you each quarter. MLPs were designed with the individual investor in mind.

These types of firms—ones that pay distributions, and have strong free cash flow with little debt—also tend to have two lower volatility and steadily increasing distributions.

These are desirable sweet spots for an income portfolio. It tends to change less in value and offers an increasing income stream. You will not hit homeruns, but you will keep pace with inflation. For example, much of the income from midstream MLPs (pipelines. etc.) is regulated by law. FERC sets the toll on the pipeline traffic. As with utilities, the regulations work to the investing public's advantage at each end of the spectrum. As a utility, consumer costs are tempered over time. As an investor, the fee increases pass through to you in the form of dividend increases.

If you seek income from your portfolio, MLPs (as well as energy firms and utility firms in general) can be rewarding. They are not without risk. They can be complex. They were designed for the wealthy, astute investor. The retail market is now exploiting them. Mutual funds, exchange traded funds (ETFs), exchange traded notes (ETNs) and general partnerships (GPs) are creating investment products wholly encapsulating MLPs. The wise individual investor can still find good income with a modicum of safety and lower volatility from the individual securities.

You can be hurt. The point of investing is to accept risk commensurate with potential reward. You expect compensation for potential loss of principal. Because you can lose, you should always employ a simple tool: the stop loss. If you buy a stock for $20 a share and it drops to $18, sell it and walk away. You will always have a stable full of other thoroughbreds paying comfortable dividends from which to replace the loser. Apply the rule, avoid the loss. Never fall in love with a stock.

Which industries create distribution income?

- Exploration and production
- Midstream—pipelines, compression, and refining
- Propane and refined fuels
- Marine transportation
- Coal and other minerals
- Alternative energy (enabling legislation 2008)

You can explore which firms actually pay dividends in this arena by visiting the group's website or the appendix. There is a significant breadth and depth of investment choice in the industries. 80 percent of retail investors in MLPs are over 50 and own these for the income distributions.

These distributions have significant tax advantages. They are complex and best suited for a well-informed investor. If you do not understand the tax issues associated with MLPs, you can still acquire these income generators in the form of mutual funds, ETFs,

ETNs, and direct investments (non-traded vehicles). In exchange for a lower distribution, you simplify the accounting process. Your ownership of individual securities will involve working with your tax consultant. Your financial advisor should have more than a working knowledge of these investments.

The accounting creates "peculiarities" for your MLP investments. Depletion and depreciation, as well as leverage, allow very significant tax deductions. The majority of your annual distribution is treated as "tax free." It is viewed as a return of principal because of these capital deductions. If you own rental property or invest in REITs, you are familiar with the approach and the tax results.

You must keep track each year of these deductions. They reduce your principal investment - your cost basis. If you sell the investment, you must use the adjusted cost basis, arrived at from these calculations, to determine your capital gain or loss. If you own these long enough to adjust your cost basis to zero, your tax advantaged income becomes ordinary income. If you die and pass these investments on to your heirs, this accounting information is ignored, and the value of the investment on date of death (or six months thereafter) is your new cost basis.

The use of tax-complex investments such as MLPs in a tax-advantaged account—your IRA, pension or 401-k plan—ignores these tax complexities. As long as your investment in these vehicles is not substantial,[89] they can be a viable alternative income investment. The issue is unrelated business taxable income (UBTI). If this figure is greater than $1,000 each year, as reported on the K-1, the plan must report this as taxable income. If you place MLPs into your retirement account, you do lose their tax accounting advantages. If you're gonna play, play well. These investments make little sense for the Charitable Trust, pension, and institutional investor for this very reason. UBTI is like a curse to them. Hence, the market for MLPs has been the individual investor.

Newly designed investment products strip the tax advantages from the core investment in the hope of playing to the larger crowd. Wall Street will always find a way to force you to "pay to play." If 80 percent of the MLP marketplace is the individual investing community, investment bankers will figure out how to sell them a product. These products are ETFs, ETNs, and mutual funds. They are costly, and reduce the distributable income by their fees. They often use leverage to enhance their size, fees, and distribution. They are unable to move as nimbly as you, the individual investor, can. Should you have to exit your position, you can do so with a fair degree of liquidity and timeliness. These investment products are lumbering beasts. Their value can be slaughtered by systemic market changes (recession, tax changes, liquidity issues, international turmoil).

You can make a wise decision or have a simple decision made for you—your call. As Warren Buffet said, "If you have been playing for half an hour and don't know who the patsy is at the table, you are the patsy." There are plenty of resources to enable you to make informed investment decisions. Start with these websites:

Naptp.org
Quantumonline.com
Finviz.com
The individual website of each company under consideraiton

Observe before you invest. Watch the markets for a few weeks. Follow websites that link from these starting points. Watch the table and the players. This is neither particularly easy nor overly difficult.

This small quadrant of the sky is vibrant with income possibility. Explore it. If appropriate, use it. Be aware of the tax complexity. Explain to your CPA that you will be using more complex tax investments such as MLPs. He will appreciate the heads-up before tax season, rather than during—never surprise a numbers guy! Most important, as investments these tools have an order to their universe. You must learn this order. Learn to play this game by the rules. That is how you win. This is not about hitting homeruns with a quick stock pick. This is about investing for your retirement income—for years.

MARKET CHALLENGES

Every country I would go to, even if it was just on a modeling job, I would go to their markets. If I went to Morocco for 'Elle' magazine, I would be in the spice markets during my off time and just come back with a suitcase full of stuff that I really wanted to try.

—Padma Lakshmi

We have seen the demand/supply/price equation at its best—and at its worst. External events forced aberrations; Katrina forced the price of natural gas up to $15. It has since dropped as low as $2. Capital markets like to work inside the fence. Some companies are riskier asset plays: alternative energy, LNG, and refineries work at the industry margins. Other risk based firms might include oil recovery and water-free fracing. Most individual investors want steady cash flow in support of real and consistent dividends. The energy markets cover the gamut of investment choice, from high risk E&P to low risk consumer product producers. These firms work in these markets, sometimes to their advantage, sometimes at severe distress.

Let's examine the various capital markets in the energy arena. Markets in the energy world can be broken down into:

- Capital
- Commodity
- Exploration and Production (upstream)
- Distribution (mid-stream) and Processing (downstream)

The **capital market** is governmental, private, institutional, and public. Government capital investment is made by tax-based mandate, captive capital, and politically directed. Private capital derives from accumulated wealth of successful individuals. It may be invested in small startups or massive scale investments; both are done on the terms of the private investor. Institutional capital is pension, life insurance. It is long term in scope. It tends to lower risk, income oriented investing rather than E&P. Public capital is just that: it typically is organized and directed by investment banks for offer to the public. Lets focus on government capital.

Government capital may be evidenced by the Chinese efforts to penetrate North American markets with capital surges for such firms as **Nexen.** The Chinese state-owned enterprise (SOE), **CNOOC**, bid a 60 percent premium for ownership. In exchange it would receive the vast Canadian assets, Gulf working assets, access to top-drawer North American drilling technology, and priority in the Canadian tar sands play.

Virtually all of the crude reserves globally are controlled by political regimes rather than private entities. While the functional differences may be debated, the caudillo "strong man" attitude is similar: These are our assets, and we will drive them as we choose. Capital investment here is made by tax-based mandate, captive capital, and political directive.

Let's look at three examples of politics affecting the energy markets: Germany, China, and the U.S. Neither of the first two is a cultural, political, or religious "enemy" of America. They are our competitors, certainly, but they harbor no desire to destroy our nation. The policies they make impact their own energy markets in a variety of ways, none of which appears to make sense. Observe.

Germany

Germany has a master plan to rule the world—of energy, at least. It is forcing carbon emission reductions—greenhouse gas emissions (GGE)—down the obliging throats of its public and industry. To do so, it has used the Japanese tsunami disaster of 2011 as an excuse to shut down all nuclear power plants well ahead of the planned 2022 phased termination. It has ordered the GGE goal of a 40 percent reduction from 1990 levels. It has implemented plans to install "white blade fields" of windmills, whose masts tower over the northern shoreline, altering the landscape, and killing myriad birds that collide with them.

This **Energiewende** (Germany's energy revolution using sustainable energy) forces the construction of even more "alternative" energy projects, whose intermittent power generation harangues neighboring nations with massive power dumps of useless solar and wind power at exactly the wrong times, at exorbitant prices. German consumers' electricity bills have doubled and will increase by another 70 percent by 2025, according to the Karlsruhe Institute of Technology.[90] This decade's need for new transmission lines alone exceeds 5,200 miles. The four main power generation firms realize that the macro goal of this renewable energy strategy is to break their backs. It offers a 20-year guaranteed "feed-in tariff" to any and all power generators: your roof, your garden, your car—bring them in and get a free tariff. Energy cooperatives dominate new power generation. While the anarchy of self-imposed power generation from your rooftop may be intellectually stimulating, it is depressing an already moribund energy market in an economy that can scarcely afford more capital constraint.

All of this "alternative" energy production destruction and re-creation is at the hands of a modern socialist government trying its best to meet the dictates of its citizenry. Top-down management in the hands of neo-capitalists may sound right in the beerhouse and the parliament. It has yet to bring about any decline in GGE. In fact, the opposite is occurring, as coal production continues to be the fuel of choice across the continent. The unseen hand of the marketplace would appear once again to trump the *magister ludis* of Berlin and Strasbourg.

China

China has seemingly limitless energy needs. It is designing enormous energy production capacity, often by reverse engineer-ing from a joint venture fool from the West. (No harm there; in the 1950s they were called **oil scouts.** In the author's county there is a valley, laid out by a watercourse wending to the sea. Each of the valley walls carry oil rigs run by two different energy firms during and after World War II. The oil scouts (one is a client) would sit on the ridges and count the number of joints drilled each day to determine well depths. They would view the trucks running up to the rigs to figure out what supplies were brought in, which told them what success was or was not on the horizon. The difference between these scouts and the tech thief of today is one of increment rather than ideology.)

To meet its projected needs, China is building out infrastruc-ture massively while buying raw materials on a global scale. It is constructing entire new cities in the belief that, if they build it, citizens will come. Their entire approach to energy policy is one of top-down authority. The nation today is the global leader in both energy use and GGE, which is no small feat. Politics bends to the will of the People's Party. This means they will buy anything from anywhere for any price, as long as it meets the master plan. Call it mercantilism gone mad.

Whether these projected needs materialize is a demographic bet. You see, their population has stopped growing in 2012. It has begun to decline. The "one child" policy has had a horrific impact. Tens of millions of female infants have been destroyed. Boys far outnumber girls today. Birth rates have dropped well below the replacement figure of 2.1 per couple. Young people are marrying later. The excess of men to women puts a strain on the 'pair bonding' process. By 2050, the population may well have shrunk by as much as 20%. While acceptable from the global population perspective, the market view is distinctly different - and diffident.

The recession in China today will play out its course. The real question is how this lack of future demand will impact this culture's "empty nation" prospects. Of course, stabilizing a country with a 1.2 billion population stills implies significant consumer demand, both pent-up and future. The end result of their population control policy is demographic: Oldsters expect their offspring to take care of them. Savings rates of 40 percent are there for a reason— health care and pension benefits are "observed more in the breach than the contract." People know this, so they plan by saving. They expect the ancient tradition to hold: You take care of your family. Does this bode well, or ill, for energy demands in the future China?

Energy policy hardly reflects the individual and family behaviors of the Chinese citizen. As in Germany, top-down decisions have a way of snaking back around to cause unexpected bites to the posterior. The massive capital dumps by the Chinese government and its SOEs have come from cultural savings habits of tens of millions. If these savings evaporate, what will happen to energy demand? If the population requires more energy over the next generation and that demand is met, but ultimately has a far smaller energy demand longer term, what will happen to the sunk capital of the citizenry? Will corruption and stagnation erode any economic growth throughout the developmental stages of this still-emerging economy? Political mandarins of all nations have a nearly perfect record in management—that is, nearly perfect failure.

United States

Finally, we should look at our own glass house. It is a house in disorder. On the one hand, the alternative energy policies of the federal and of many state governments dictate new energy sources and new means of power generation. On the other hand, like drunken sailors, investors sink billions in capital into E&P, distribution, gathering, and processing, as well as into further development of energy infrastructure needs. MLPs are growing at an extraordinary rate.

Regulatory requirements act as a governor on the capital machine. They are necessary and often sufficient. Yet, rules are entirely different at this policy level. Murkiness is a given, as is change. Even the biggest capital players—banks (institutional) and the oil majors (public)—have been seriously hurt here. The challenge is essentially one of information. The reader has to maintain awareness of the politics of changing fortunes at the national and global decision-making levels.

The issues of energy production and distribution are poorly managed from above. The markets tend to a boom or bust cycle. We may be witnessing just such an expanding bubble in natural gas prices today. The dichotomy between markets and policy makers is constantly being bridged by brave participants, whose bridges are often destroyed by weaker opponents to cooperation. Opportunity should take precedence over safety.

Commodity markets are composed of two interrelated fields: raw materials and the trading of them. The raw materials market is a complex of producers, refiners and delivery agents. These are regional, national and global. This market attempts to respond to demand with pricing tools. These tools are the daily events on the trading floors of the commodity exchanges in Chicago, New York and across the world. The tools use numbers, the price for a Btu of natural gas or a barrel of oil, expressed in dollars per unit.. The numbers express the moment by moment demand and supply for the raw commodity. These two aspects of the commodity markets – material and price – can be influenced by exogenous events, events outside of the markets. Politics, weather and economics are exogenous. Capital flows within the commodity markets is enormous, far greater than the value of the raw materials themselves. The matrix of capital flow feeds the equity beasts.

These are the third and fourth levels of energy capital. **Exploration and production** companies, E&P, are the upstream capital beasts. Jake works here. Fracing is at the heart of their efforts. Whether for gas, oil or NGLs, these firms drive the extraction of hydrocarbon. They tend to consume large amounts of capital at regular intervals. **Linn Energy** is an example. It has grown organically through the acquisition of fields and their subsequent development with hydraulic fracturing.

As capital volume (investment capital) increases in the exploration and production world, risks increase for the capital market. The capacity to absorb more and more capital is beginning to exceed much of the delivery and demand characteristics of the current American energy market. Oil demand declines because of slower economic activity and higher fuel efficiencies for autos. Current production offsets imports, which lowers overall costs and promotes further capital investment. The spreads between U.S. and global prices for crude oil give the U.S. producer the advantage. Internal spreads across the U. S. markets offer price discounts to Midwestern producers with lower overhead and to eastern producers with closer access to demand markets. Higher well costs deter gas E&P, resulting in lower rig counts at dry gas sites. Rigs move to the oil fields. Oil rig counts offshore are finally increasing after the Macondo moratorium dictated by worrisome federal regulators.

Midstream firms are the pipelines of the previous chapter, the regulated delivery lines whose invisible arteries will expand so significantly during the next decade. Capital intensive initially, they become capital distributors once complete. Like a bridge or a tunnel, their capital value lies in their extended use. **Downstream companies** take the distribution and turn it into a final product, the plastics, fertilizers and fuels we accept without question or awareness. **Calumet** is an example of the downstream result. From WD-40 to Turtle Wax and literally a thousand other consumer products, they turn the distributed gas and oil into the indispensable items in every kitchen and garage. The capital flows to these firms tends to be slower, in line with their expansion of product into market. It is as much debt as equity. The capital demands are reasonably consistent and flex with the economy. These demands are far removed from the government, commodity and upstream markets and their concomitant capital requirements.

Each of these complex capital and policy events impacts the other in a dance macabre. Capital chases return, is met by risk, and given a stern glance by regulatory matrons. Competitors take the hand of the nearest dancer. The energy demand music continues, but at a lower volume. When is the punch bowl removed? Who takes it? Will new dancers appear to occupy the floor?

Current Capital Opportunities

North American distribution channels are the continent's cardiovascular system. Processing and storage are the major organs of the national energy body. The United States enjoys a century of experience, technology, and knowhow across the entire spectrum of the energy industry, permitting investors a wide variety of capital choices and access to a deep array of financial and geological information. This lowers perceived risk in the capital market and draws new funds into play. This is where the capital action has been for the previous five years, the action for income. Investors look to receive a comfortable income stream from taking risk on the build-out of national infrastructure.

Distribution, storage, and processing are the ground ball plays in the midstream ballpark. In addition, the unique tax structure of the MLP format that's typical in this play effectively lowers the cost of capital for investors because of its pass-through nature. The lack of direct commodity market price exposure reduces perceived risk from raw materials pricing pressure—in either direction. Judgment calls from a capital perspective are driven by actual demand for transportation, gathering, and processing, rather than from such androgynous financial statement figures as P/Es or FCF. These are real, demand-driven investments with significant and immediate cash-flow payoffs. While the capital is sheltered by depreciation, the return is significant and immediate "cash on cash." Revenue is fee-based and mandated by FERC; it is rock-solid.

Capital appreciation occurs through organic growth and acquisitions. The organic growth comes from infrastructure build-outs: pipelines, processing, and storage. The acquisitions are from asset spinoffs by the oil majors, and wise pickoffs of the assets from excessively debt-laden competitors and utilities. New equity is issued for their purchases, offering fresh capital to an intensely capital-driven market. Pipelines are extremely capital-intensive projects, as much as a million dollars per installed mile, depending on both physical and demographic terrain.

Once the capital is "sunk," it works through the dividend process at rewarding the risk-aware investor with steady cash flow. The growth of dividends each year adds value to the proposition. Long-term investors are rabid fans of an investment that pays steady dividends that steadily increase. Financing risk declines as capital risk is managed well by the infrastructure firms. Even when the systemic risks of 2008-09 hit all markets, these firms, while cut off from new capital, continued to pay out their cash flow to shareholders. Neither markets nor investors forget these acts of fiscal integrity. The harmonious relationship between capital, market, and investor is complete.

All of American industry relies on these capital investments and their successful conclusion. Pipelines distribute fuels for cars, trucks, and planes. They send methane to utilities for electricity production. Processing plants convert base liquids to fuels, additives, and feedstock. You get that new can of WD-40. Imports of crude have dropped dramatically, by 40 percent since 2005, from the Middle East and South America. Exports of kerosene and diesel are exceeding levels not seen since the 1950s. LNG import terminals are being licensed for turnaround to export.

PUBLIC AWARENESS CHALLENGES

Honesty is the best policy - when there is money in it.

—Mark Twain

These issues face the industry, the public, knowledge brokers, power brokers, regulators and politicians across every layer of American society today. Many in the industry and its regulators have been "manning up" since the development of hydraulic fracturing, seismic technology advances and horizontal drilling. Some have ignored them. The local public has had to become aware as the landsmen and rigs have rolled into towns. Public groups such as NGOs and the media have taken an interest in the past two years. Politicians blow with the wind. These concerns are at the forefront of the challenge of unconventional gas and oil. Technology, regulation, applied science, legal and societal questions now face the industry and the nation. Your interest in these issues has brought you this far. Prepare to go deeper, much deeper, into controversy. Our point here is less to take a stand than to make you aware – so that you can make that stand. Your read; your call.

Industry Issues

The National Commission on the Deepwater Horizon Oil Spill and Offshore Drilling set out the need for sweeping reforms that would accomplish no less than a fundamental transformation of the oil and gas industry's safety culture.[91] Many of these reforms are regulatory; some are industry-sponsored. All are impactful on the energy exploration and production from the Gulf. These reforms will be costly to the offshore oil and gas industry. As for blame, if you ask riggers in the field, they first implicate the engineers and management rather than the regulators. The field knows what the problems were, how they were ignored, and what the results were and will always be: a disaster for all concerned. Those in the field, from roustabouts to geophysicists, know that excess meddling by unqualified management seeking faster production schedules are at fault. The disaster was one of management. The Gulf cleaned itself up faster than the offending management teams. Fortunately, HSE (health, safety, and environment) training has risen to the challenge. In the field and in the lab, safety remains paramount.

The changes being brought about by the invisible hand of the marketplace are numerous. For example, the many issues around water use arising today from hydraulic fracturing are being addressed, if not yet resolved, by the marketplace. Recycling, water-use substitution, and gel fracing are important examples. Water disposal into injection wells has been a common solution for decades; now, it has been upgraded to EPA-approved standards. Sometimes, it takes an earthquake to raise up a village. The oil and gas village is rising up with many solutions.

Green fluids are being developed for offshore and arctic environments that can be reapplied to local fracing in the Marcellus, Bakken, and other fields. The EPA has approved a biocide that has low overall toxicity and easily breaks down in the environment. It protects the gel agent (the guar from your chewing gum) from breaking down and reduces downhole corrosion. Flowback from the frac site is beginning to be recycled. In British Columbia, this is accomplishing 50 percent reuse rates and reducing shipments of water by a similar amount.[92]

Public Awareness

We often hear about the lack of a level playing field for renewables. How level is a field so tilted with tax support to alternative energy sources? How fair is it that more than half a dozen solar firms received as much as $1 billion of federal tax support, and then went bankrupt? One could easily argue that the federal regulators have acted imperiously in artificially backing certain companies and industries at the expense of other energy-production sources. This kiting, "hand on the scale" of the market has had the result of negatively impacting the very companies and industries for which advantage was sought. At last count, seven U.S.-based solar firms that have been subsidized by federal handouts of taxpayer assets have failed. It's your choice. Do you prefer the visible, heavy hand of government intrusion and reverse taxation, or the invisible hand of the marketplace?

While European dictates from Strasbourg demand ever less carbon-emitting energy production, their unintended consequences have actually resulted in an increase. Coal is their tax penalty-box answer to the carbon conundrum. Irrespective of their decrees and dressings down, European "carbon emissions" continue to grow. Meanwhile, here at home, despite our personal disregard for carbon concern (outside of the government), the nation's carbon production declines. Markets have driven down the price of "carbon credits" to such lows that they no longer have any value.

Natural gas stands accused of powering the nation more efficiently. According to a September 2012 commentary in the academic journal "Climate Change":

> Natural gas is widely considered to be an environmentally cleaner fuel than coal because it does not produce detrimental by-products such as sulfur, mercury, ash and particulates and because it provides twice the energy per unit of weight with half the carbon footprint during combustion.[93]

The use of natural gas for home heating, hot water for cooking and cleaning, and for electricity generation is energy efficient, environmentally beneficial, and an inexpensive choice for every family in America. The source of this gas is hydraulic fracturing and horizontal drilling. Done wisely, as regulated by the states, this production and distribution will continue to drive demand and supply. We have a 100-year supply of natural gas for the nation at today's usage levels. In fact, we need more than 250,000 miles of pipeline installed during the next two decades simply to keep up with the delivery of new gas supplies from the deep Earth.[94] The 250,000 miles of pipelines crisscrossing the nation today must continue to expand. That means jobs, tax revenue, and an enhanced appreciation of a safe and clean environment.

Natural gas consumption results in significantly less "carbon emissions." Setting aside for the moment the meaning of this odd concept, the result should enthuse our environmentally concerned fellow citizens. Natural gas carbon emissions are 44 percent lower than coal and 25 percent lower than gasoline. Are the environmental lobbyists celebratory? Do they congratulate the energy industry for meeting and far exceeding their demands? The answer will surprise you—and it should shame them.

The recent policy statement from an environmental organization represents one response:

> Fossil fuels have no part in America's energy future – coal, oil, and natural gas are literally poisoning us. The emergence of natural gas as a significant part of our energy mix is particularly frightening[95].

It goes on to say that fracturing is an abomination, and they will pull no punches in halting its use because its production may harm the environment. More to their point, its use interferes with the nation's conversion to alternative fuels. In addition, there may not be as much gas as we have been told, and its extraction most certainly, probably, might harm us. If there is a finite amount in the country, then exporting it is foolish. We should keep what we have for the future and plan for its use as a transition fuel, a fuel that will allow us to become entirely hydrocarbon free. Jobs are at stake. We need to stop producing natural gas today.

This comes after a decade of support for the idea—and $24 million in financial support from the energy industry to the likes of such as this NGO. So much for objective science. Let the games begin.

Fracing has captured the attention of many in media. The attention is less on detail than storyline. Respelling frac with a "k" is not the only liberty they've taken in their reports, which lack balance, as a 2011 study shows.[96] This report compares media content and attention to the work and research at Cornell [97] and Carnegie Mellon.[98] Both research groups focused on the GGE footprint from shale gas lifecycle emissions. The Cornell report found these emissions exceeded those of coal; the Carnegie Mellon found no such evidence. The former received 24 supportive stories, while the latter garnered but one local newspaper mention. We can neither ignore nor repudiate this simple fact: Sensationalism sells. Reading on this subject requires a deeper analysis of the press, with an understanding of biases both conscious and unconscious.

The most recent issue of the academic journal "Climate Change" contains a detailed scientific refutation of the first Cornell report—from research performed by yet another set of Cornell scientists.[99] It details challenges to the process, tools, and results claimed by the first report. A fourth report from an international think tank[100] demonstrates similar conclusions. How many of us will read all four reports to develop a clearer understanding of the complex issues? What does the typical reader know of "life cycle systems analysis," GGE, "time intervals," "atmospheric residence time," "heat content vs. power generation potency" or GHG/GWP? This discussion is important. Who is willing to investigate? Who will rely on the press to summarize? Read primary sources for yourself and arrive at your own conclusions. They are actually quite fascinating.

That's four reports in 18 months on the same subject, from university-based investigators having ties to *neither* the energy industry nor the environmental industry. The score is 3 to 1 in favor of significantly reduced GGE resulting from the substitution of shale gas for coal in electricity power generation. Even the EPA concluded that the first Carnegie report is flawed in substance and in style.[101]

Macondo was a horrible event that made international headlines; yet it remains a local event, caused by personnel, managerial, and systems errors compounded by one another. The natural effects on the ecosystem of the Gulf of Mexico have surprised everyone involved. The massive spill has all but dissipated in a relatively short time frame, with oyster beds back in production and dolphins, fish, and seabirds at or near their usual populations. It would appear that Gaia is quite capable of cleaning up our messes, once left to her own devices. The mysterious processes of natural cleanup on the floor of the Gulf will be studied for a decade.

The learning has been enormous. The industry itself has taken the responsibility to form two independent response teams, in light of their own and government findings. This is in addition to the new federal mandates for HSE. These teams are funded with more than $1 billion in reserve engineering assets and personnel. The environmental results have surprised everyone. Bacteria at the bottom of the sea have apparently consumed the entire amount of petroleum released from the blowout. Public opinion and the markets have consumed the entire leadership at BP: management processes and systems have been altered radically.

The lesson from Macondo, like that of Exxon Valdez, should be taken as a sobering response to a nightmare. The response was from all members of society, and will affect each of us. Wise responses allow the exploration and production of unconventional gas resources to continue apace, to grow and to dominate the energy generation story in the U.S. Danger and opportunity are often "inside straights." Their partnership is crafty, necessary, and useful. We should look upon these disasters as stepping-stones for the future as opposed to roadblocks to progress. The myriad rigs in the Gulf of Mexico withstand the hurricanes of nature and political verve. The industry and its regulators should each stand proud of their constant efforts to monitor and upgrade the most dangerous process on Earth: the extraction of hydrocarbons from deep beneath the ocean surface, even deeper beneath the seabed. NGOs, legal teams and regulators have each stood forward, often against the wind from the industry, to urge, to force change upon an often reluctant group of egos. We should all be proud of the work done in the field, the laboratory, and the offices of the industry and its regulators.

Regulatory Challenges

Managing environmental risks while contributing to scientific advancement - all the while increasing profitability - are key components of the long-term goals and policies for many in the oil and gas industry. No one wins from a Macondo-style blowout. Everyone wins from a safe and profitable drilling operation. Images of evil corporate bosses stealing millions from poor farmers and ranchers are fantasies designed to sell tickets. Frankly, there is more money to be made by playing by the rules. There will always be evil corporations; they will always be the exception that proves the rule. The rule is: do well by doing good.

Communication between drilling crewmembers, the local community, and state regulators is integral to alleviating concerns expressed, and to sharing processes across disciplines. Open communication channels are the key to properly functioning sites.

State authorities have primary jurisdiction over much of the hydraulic-fracturing industry. If each state's legal and regulatory structure is at least as stringent as that at the national level, then it retains final arbitration on any legal issue. This is a source of great pride at the state level—and of much consternation at the federal and NGO level.

Some wish to use the Clean Water Act (CWA) as an example of why a federal mandate for drilling supervision and blanket federal regulations would impose better supervision. They dislike the "Halliburton exemption" of the CWA. These observations appear to be political in nature. Politicians and voters will have to decide whether state regulators or federal ones are better at the task of supervision. Local authorities do seem to know the land and its resources better than carpetbaggers coming down from the North to impose their justice on unwary citizens.

While the Bureau of Land Management (BLM) and the U. S. Forest Service oversee developments in the federal lands, national regulations for the protection of all aspects of the environment are reflected in state regulations. The application of and subscription to regulations are monitored by state regulatory agencies, which implement many federal laws under mirroring arrangements. As an example, the Clean Water Act grants primacy to the states to regulate water discharges from pads, surfaces, injections, emissions, and ultimately to ensure land reclamation.

State application of regulatory oversight is today presumed to be far more efficient than a federal jurisdiction. Local regulators have local experience and knowledge. They apply their version of federal mandates according to best practices in the field. Local assessments are often more demanding and can claim greater jurisdictional application to public and private land holdings. Area professionals are more familiar with variations in geology, hydrology, climate, topography, and law. Local and regional economic, industrial, and population trends are most effectively dealt with by state regulations.

Active environmental management programs by the major companies in shale gas development are cost-effective strategies. Pad sites are cleaned and returned to their natural condition, or better, once the annoying drilling is complete. What remains is an unadorned "Christmas tree" and perhaps a small storage bunker. Pipelines disappear beneath the soil, which is replanted, or cross rivers, roadways, and mountains in an invisible web of strength. Well-worn roadways are repaired or replaced—or should be.

Water management is an economic and social decision. Its use is both costly and controversial. Frac fluids today are being designed to reduce reliance on water. Recycling and reclamation of down-bore water is becoming universal. Of the water used, as much as 60 percent comes from the deep field. Less than 30 percent returns to the surface as a salty liquid (brine). Of this amount, 40 percent is recycled and reused. The disposal and treatment of this brine is an expense to the bottom line, and well-observed. Substitution of gels, carbon dioxide (in the form of carbon sequestration, the environmentalists' favorite GGE tool), or LPG are each reducing water use in large volumes for many new drillings. EPA-approved injection-well technology and well use is increasing. Water use and reclamation has become the newest "profit center" for many E&P companies. Water pipelines reduce trucking and the concurrent emissions.

"Prevention works," as Benjamin Franklin coyly observed in his Pennsylvania homeland. As E&P firms prevent excess water use, they work for the betterment of the community. They do so while earning a profit. Industrial practices attempt to minimize and eliminate negative health effects to field workers and local communities. These behaviors are constantly being adapted and modified.

Environmental Challenges

Virtually all environmental concerns raised on the issue of unconventional shale gas and oil development center on local concerns: the rigs, pads, equipment, facilities, trucks, roads, pipelines, and storage facilities. The only "global concerns" have to do with potential seismic events, and the ever-present, ever-renamed "climate change." The latter is clearly addressed by the facts we have repeated: America's GGEs have declined significantly during the exploitation of unconventional oil and gas reserves. No other nation has seen such a dramatic decrease over any time period, much less one as short as seven years.

The seismic issue has been raised as a result of several incidents, one in the UK and the others in Arkansas, Ohio, and Oklahoma.[102] Evidence compiled by the National Research Council and the U.S. Geological Survey indicates that microquakes of 3.0 or less can be caused by **injection wells** when these are poorly placed. Recently released research results from Shawn Maxwell, at the Society of Petroleum Engineers workshop "Injection-induced Seismicity" in August 2012, demonstrates that several elements need to come together to cause these micro-quakes: a weak fault zone; one that must be under stress parallel to the fault; the liquids must parallel this zone; and they have to enter the zone faster than any pressure relief that may be available.[103]

Calling for a moratorium on production appears entirely unwarranted. Correlation is not causation. Injection wells are a subset of unconventional oil and gas wells. Their use is rather limited and always under EPA regulations. It would be as if the city of Long Beach banned oil drilling because the city finds itself slowly sinking, relative to the surrounding land. Yes, it could be banned, but why?

The issues here are far less complex or overriding than elsewhere in the world of human impact on the environment. Each of these issues is localized, rather than globally systemic or pandemic. Concerns raised are clearly expressed in many instances. These specific concerns are real, yet each can be engineered. Applied technology solutions have an interesting impact: They tend to solve more than one problem. Let's itemize and review the environmental concerns:

- Pad construction and operation
- Wellbore development and maintenance
- Injection processes, effluents, and emergents
- H2O sourcing, use, waste disposal, and treatment
- Atmospheric emissions during and after wellbore
- Health effects on workers and local citizens

Pad Construction and Operation

A well pad today with horizontal drilling can access the same deep rock field of as many as 16 vertical wells. The reduction in roads, trucking, pipelines, storage, and production facilities is clearly advantageous to the surrounding grounds, habitat, and local environment, as well as to the public. The very process of horizontal drilling is friendlier to the locale than previous drilling techniques. Yes, they are noisy, dusty, and visually disturbing; yes, they disappear in a four to six weeks.

Each site itself must be properly prepped with a functional yet clean and cost-effective design. Roads and sites are subject to state approvals that vary between regions. Materials, access, and use as the pad is under construction should meet local standards, whether rural, agrarian, or suburban. Site development includes design, road and erosion planning, clearing, excavating, utilities and pipeline installation, backflow control and contamination abatement, soil and seed management, materials use, and the permitting, approval, and supervision of each. Ultimately, the site is, or should be, returned to the owner's use in better condition than at acquisition.

Habitat disruption or fragmentation can arise. Accidents during construction or operations must be monitored and reported. Remediation is the responsibility of the site user. This should go without saying, yet some users try to ignore their clear responsibilities. Hence, the clear case for regulatory oversight, as expressed in Pennsylvania.

The pad itself may be a few acres, and under intensive use for a few weeks. The drill site is chosen for its ability to drain as much gas, NGLs, and/or oil as possible, with as few fracs as necessary. The drilling platform area is cleared and leveled, and then a thick polyester mat is laid beneath the well pad. Under the rig itself lies a thick rubber pad. The drilling area is trenched to further contain site events. All materials used are stored on their own containment platforms; diesel fuel tanks, for example, have as many as five containment resources. Berms surround the site to direct rain and snow fall to sump pumps that hold the water for recycling into the well, as directed by regulations.

Wellbore Development and Maintenance

Drilling mud handling, use, and disposal may involve water-based muds with additives or oil-based muds with diesel or other fuels. Closed-loop recycling removes cuttings and recycles the water back downhole.

Before the drilling begins, a special conductor casing is drilled into the site, the first of seven to nine bore casings used to attempt to completely isolate the wellbore from the surrounding earth. The first 800 feet or more of the well is drilled with an **air bit** to avoid contamination of the porous, often water-soaked, ground. The remaining six casings are steel tubing and cement. These stabilize the wellbore as well as protect the land and water. Once the well depth is reached, the pipe is tripped out and production casing is run in, and then the pipe is replaced.

The vertical wellbore, once complete, is secured in place with special fly ash-based cement (burnt coal) which is **mangled** down the pipe outer bore, or casing. This is meant to be an impermeable seal between the well and its surroundings—such as the local aquifer. Virtually all documented cases of well breach occur because of a poor cement job. Industry records on new well-casing failures, and annual state reports of new wells, indicate that six percent of new well casings fail, according to www.DCS.org.[105] These failures are eliminated with multiple casings, wisely drilled.

The water from the drilling, the cooling, and tailings mud cannot be allowed to reenter the local aquifer, so it may be transported to injection wells. It may be recycled several times. Cuttings are disposed of at landfills, as required by state law. The reserve pit holds the recycled water, stripped of cuttings by the shaker, awaiting reuse. Failure of the reserve pit is a serious concern for the crew and the local community, such as the agricultural areas of the Marcellus or the Fayetteville fields.

The blow-out preventer, or BOP, does as its name suggests. In the case of a pressure flow problem, it terminates access to the borehole. Drilling team members and state regulators test it frequently.

Injection Processes, Effluents, and Emergents

Much concern has been expressed from a variety of sources regarding the chemicals used during the hydraulic fracturing process. Many have said these chemicals are secret formulas, private labels, and another example of devious behavior by corporate muggers. In fact, most wells—and the chemicals and proppants used in each—can be thoroughly researched online.[106]

The Fayetteville fields in Arkansas are experimenting with an eco-friendly frac fluid manufactured **by Echosphere Technologies.** Their Ozonix™ treatment eliminates the use of potentially harmful biocides down bore. As of July 2010, 40 wells had used the product successfully.

Water Sourcing, Use, Waste Disposal, and Treatment

Water use at a hydraulic fracturing well site is less than a Florida golf course uses in three weeks. It uses less water than most other industries, and far less than other energy sources. Quantity is not the issue. If you dare to compare the water use by a chip manufacturer to that of the oil and gas industry, chip manufacturers would be disgraced; their use is 10 times that of the oil and gas industry. Local sourcing use and reuse is more important.

Sites are now developed that attempt to make use of local rainwater by damming temporary rainfills into small reservoirs to capture the rainwater. Berms enclose most well sites, and rain or snow runoff is collected for use. Local water sources are pipelined in or trucked. While water usage per site can be one million to six million gallons, water usage for hydraulic fracturing is typically less than two percent of total human water consumption.[107] This figure can be misleading: The cattle country frac site may use five to seven percent of the water table; the site on the river may use less than one percent.

Water Use in Hydraulic Fracturing

The name says it all. Hydraulic means water use. In 2007, 21 billion gallons of produced water was used in the oil and gas industry. That amount has increased significantly, and 55 percent is re-injected for enhanced recovery (recycled); 39 percent is injected into old wells; and five percent is reused, treated, or surface disposed. Public and profit demands for enhanced improvements in water use continue to drive many new firms to explore capital-impressed opportunities.

Water use is costly. Awareness of efficient water use is both financially and socially rewarding. Frac water cost has more than doubled in 2012. Reducing that cost translates directly to the bottom line. It also lifts both gas and oil plays in the public eye. Ensuring "best practices" in HSE is rewarding—financially, personally, environmentally, and conscientiously.

Green Hunter serves as one of many fine examples, setting the mark for recycling, reuse, reclamation, and EPA-approved disposal sites and techniques. They meet or exceed the EPA requirement for Class II injection wells, today's regulatory standard. They claim to be technology agnostic, with solutions ranging from injections wells to aboveground storage to optimal water usage and treatment in frac operations. Lower truck cartage reduces emissions. Setting standards for the industry also gives horsepower to regulators to further design best field practices.

GASFRAC, from Canada, completely eliminates water use with an LPG closed system. A smaller pad footprint, reduced truck use, and no surfactants, biocides, or refractants creates higher recovery rates at lower horsepower frac spreads. The fracs are deeper, more evenly distributed, more efficient, and have greater recovery rates of hydrocarbons. Their use of propane, one of the NGLs from the frac recovery of many wet wells, completely substitutes for water in the frac-spread cycle. Their sites have significantly reduced truck traffic and emissions. Flaring is minimized through higher recovery rates. The 25% added cost can be recovered by the additional extractions.

Carbo Ceramics is reducing frac time and bad sand use (Chinese sand) to bring about greater productivity with their high-conductivity ceramics. Improved IP is a direct and immediate result. The average EPA leakage report results in costs in excess of a quarter of a million dollars. Their technology can reduce leakage potential while using lower volumes of proppant and water. Their use of northern white sand, the highest grade available today, affords their ceramic proppants higher recovery and estimated ultimate recovery (EUR) rates. They minimize crush, increase thermal resistance, and offer uniform sizing. This results in less sand and proppant use by volume, reducing truck traffic and emissions. The cost savings affect both profitability and public perception.

Ridgeline recycles water for the oil and gas industry through electro-catalytic treatment. They separate water from its contaminants by ion exchange, removing total dissolved solids (TDS), heavy metals, and chlorides in a cost-effective manner. They claim lower cost and better results than cavitation, evaporation/distillation, reverse osmosis, or electrocoagulation techniques. Reduced truckage impact is a direct result.

Fountain Quail Water Management also recycles onsite and has expanded its presence in the Barnett. Their ROVER ™ system is a mobile answer to water sourcing, recovery, treatment, transportation, and reuse. It can recycle 10,000 gallons daily, removing particulate and organic suspensions. The resulting brine can be mixed with propps and reused.[108]

Firms such as **NeoHydro** also clean the brine at the wellhead. They treat the water without the use of potentially dangerous biocides, while removing heavy metals and solubles. Its "electro-oxidation process" has been successful with several wells. The firm claims that the cost savings over water shipment or bioremediation can exceed 70 percent. Clearly, the savings can be substantially more than the added cost of these new treatments.[109]

Halliburton's proprietary treatment process "CleanWave"™ uses an eltrocoagulant to destabilize the brine through ionization. The suspended matter rises to the surface and is skimmed, leaving water for reuse in the well. It claims a 26,000-gallon daily capacity.[110] Their in-field activity shows significant cost reduction and environmental protection.

MBI Energy Services (MBI) of Bellfield, North Dakota is a specialty trucking firm, built on capital from Thrivent,[111] a Lutheran capital management firm. While having God on your side always makes sense, having Lutherans from Minnesota behind you is an assertion that their propinquity deserves the best. MBI delivers water and frac fluids onsite, while hauling away the saltwater for recycling. It also carries mud, sand, proppant, and oil. The service industry, a quiet subset of the oil and gas industry, is always needed and always profitable. In just a few years, MBI has become the largest private employer in North Dakota. Its HSE programs set the standard for the state and for the Bakken fields. Visit the area via web and observe for yourself the chaotic industrial turmoil ever-present, then imagine doing all this safely and profitably for the firm and for the environment.[112]

Hekkmann Corporation has just acquired Power Fuels, making it the largest environmental services company in North Dakota. They operate 1,000 water trucks, nearly four dozen disposal wells, 6,000 tanks, 200 tank cars, and 300 miles of pipeline. The firm plans to become the largest ES firm in the U.S. If you think the oil and gas industry isn't "very big" in the environmental services industry, then you had better check your spelling of HSE—the "E" stands for environmental awareness and responsibility; it is more profitable to be aware of your environment than to destroy it.

NEXT Legacy Technologies Inc., a Canadian firm, is owner of a new technology that can frac conventional and unconventional, vertical, directional, and horizontal oil and gas wells with as little as 10 gallons of water per zone. This compares with a conventional hydraulic fracturing process that uses two million to six million gallons of water and upward of 20 trucks.

All the company's fracing compounds are verified nontoxic. The "black box" technology also enables a small crew to frac one or more wells per day. NEXT technology can be used on any well, whether it's a standard vertical hole or a two-mile horizontal wellbore with 20 to 30 separate fracture zones. In short, NEXT may very well prove to be a credible rival to conventional hydraulic fracturing.

Other firms are working on casing improvements, dual synchronized weight drops for seismics, multi-staged fracs on multi-well pads, and carbon-sequestration opportunities. The contribution to environmental improvement is obvious at every level of industry, from the small operator through the midstream MLPs to the giants. HSE is an integral part of the industry. It is both wise and profitable.

Even the EPA is in on the game. It has finalized its "green completions" rule requiring drillers to install equipment at gas frac sites to capture emissions (organics such as methane) for return to the production stream. The rule will apply to gas processing plants, compressors, controllers, and storage facilities. EPA believes the cost will be offset by the sale of recaptured gas. These regulations, which take effect in 2015, are intended to reduce leakage and flaring.

Each technique influences local community impressions, reduces costs, and demonstrates the ever-changing industry HSE standards. Most are simple field applications. They reduce overhead while improving safety. The industry does have to sharpen its image in the public's eye. The stereotype of the greedy company sucking the life out of the fields of poor agricultural America has to be countered with real images of truthful actions taken at every level of unconventional gas and oil shale plays.

The industry's fortunate shale play workouts have created enormous economic opportunities, far in excess of the capital available for full implementation. Individual and business successes speak volumes to the demands for market-based, cost-effective, U.S.-centered energy production. Efficiencies are constantly evolving in the field and the labs. They raise upbore very quickly to senior management. Shell has recently acquired a 26,000 BOE production field from Chesapeake for $4.3 billion, an investment that will pay for itself in less than two years. This bold new world of shale is evolving daily, with 3-D seismic, hydraulic fracturing, and horizontal drilling spreading to new regions nationally and globally.

The most frequently claimed concern is that fracing contaminates drinking water. One claim is that fracing creates cracks in rock formations that allow chemicals to leach into sources of fresh water. The major problem with this argument is that the average shale formation is thousands of feet underground, while the average drinking well or aquifer is a few hundred feet deep. Separating the two is solid rock. This geological reality explains why EPA administrator Lisa Jackson, a determined enemy of fossil fuels, recently told Congress that there have been no "proven cases where the fracing process itself has affected water."[113]

A second charge, based on a Duke University study, claims that fracing has polluted drinking water in Pennsylvania water wells with methane gas. Methane occurs naturally and isn't by itself harmful in drinking water (though it can explode at high concentrations). Duke authors Rob Jackson and Avner Vengosh have written that their research shows "the average methane concentration to be 17 times higher in water wells located within a kilometer of active drilling sites."

They failed to note that researchers sampled a mere 68 wells across Pennsylvania and New York—where more than 20,000 water wells are drilled annually. They had no baseline data, and thus no way of knowing if methane concentrations were high prior to drilling. They also acknowledged that, for various reasons, methane was detected in 85 percent of the wells they tested, regardless of drilling operations, and that they'd found no trace of fracing fluids in any wells.

To be clear, fracing uses toxic chemicals. However, the reality is that 98 to–99.5 percent of the fluid injected into fracture rock is water and sand. The chemicals used range from the benign, such as citric acid (found in soda pop), to benzene, a known carcinogen in concentrated volumes. States like Wyoming and Pennsylvania require companies to publicly disclose their chemicals; Texas recently passed a similar law, and other states will follow.[114] The Halliburton website[115] lists each chemical used in each state where it participates in or supports the fracing process. These chemicals are listed as to their percentage downhole, their actual chemical composition, and their common use. You will discover that the chemicals are found in stain remover, liquid detergent, hair wash, cat litter, air freshener, shampoo, and of course food, ice cream, and vitamin stabilizers. These chemicals are indeed common, both in use and in substance. The website www.fracfocus.com shows many well sites in the U.S. and lists each ingredient (water, gel, sand, and chemical) that goes down the borehole. Check for your area. You will see the newest well drilled, and each component of the mix used for that well. This information is required by each state's regulatory body. No mysteries here. Only proprietary substances, designed to the shelf below, are excluded—and more firms are disclosing the chemical compounds therein, if not the actual formula.

Marcellus wells are followed at www.fractrack.org. The northeastern Pennsylvania town of Dimock is ground zero in the water controversy over hydraulic fracturing, and Houston-based **Cabot Oil & Gas** is the alleged bomber. Residents there agreed to a confidential settlement with the company following a long-running lawsuit alleging their well water supply was polluted with methane gas and toxic chemicals as a result of fracing. Cabot denied responsibility and has since been treated like a rogue outlier. At this date, the company has declared its intent to continue with its fractious approach to downhole chemicals— and the industry has backed further away from any support for its position. The legal positioning changes monthly. Guess who wins these arguments. Neither the drilling firms, nor the public – only the lawyers.

Drillers must dispose of fracing fluids, and charges are made that disposal sites also endanger drinking water, or that drillers deliberately discharge radioactive wastewater into streams. The latter accusation inspired the EPA to require that Pennsylvania test for radioactivity. States already have strict rules designed to keep wastewater from groundwater, including liners in waste pits, and drillers are subject to stiff penalties for violations. Pennsylvania's tests showed radioactivity at or below normal levels – NORM, normally occurring radioactive material.

Water use is a serious concern at any frac site. The deep, horizontal wells can use two million to six million gallons of water. As much as 60 to 80 percent of that water remains down bore, despite the extraction process. What is withdrawn from the bore is today more often recycled and reused for other drill sites. This makes economic sense to the drilling company, as water is costly to buy and haul. Much of this flowback is reused at the next site.

The transport of this wastewater by large diesel-powered trucks causes consternation to locals and the authorities. Use of pipelines for water conveyance to and from the pads is becoming more common today. These lines can be rolled, as coiled tubing is, and deployed across varying terrain, utilized for water distribution to the pad, then rolled back to the wheel and taken to the next site, much like a reeled garden hose. Barges are now being deployed for water transport in the riverine sites. These reduce truck cartage volumes by a significant margin. They have the added value of storage at the barge-docking site. These tanks will hold any contaminated and disposed water—frac water or city water—for ultimate disposal down river.

In fact, studies done by Harvard and MIT researchers show the water intensity of shale gas ranks among the lowest of all fuel sources. Across the life cycle, shale gas-fired power generation consumes only half the volume of fresh water per megawatt hour compared to coal and nuclear.[116]

Recycling of flowback can be achieved in several ways. Solids removal includes particulates and heavy metals such as aluminum, ferrous iron, and sulfur. The processes of ozonation and ionization remove suspended solids. Desalination removes salt from the brine. Total suspended solids (TSS), total dissolved solids (TDS), oil, grease, and salts are the solubles that may be removed. Salts removal is least important to the reuse of the water down bore, but valuable for ultimate human recycling. TSS and TDS are the central focus for efforts in brine water treatment. Coagulants, ionization, ozonation, ultraviolet treatment, biocides, and filtration are typically used onsite, while central treatment facilities may be employed where treated water results exceed well reuse demands, or where transportation issues are of less importance. [117]

Atmospheric Effects: Drilling, Flaming, and Trucking

Flaring has been a part of the O/G industry since gas was first brought up from a well. Use for the gas was limited, so it was burned off. These flares are hot and noisy, and can be dangerous. Are they an environmental hazard? As long as the methane is completely burned, the answer is no.

Recent research on the subject of the release of methane into the atmosphere during extraction and transportation reach differing conclusions. Three studies—one from Cornell University in Ithaca, New York; another from the EPA; and the third from NOAA/University of Colorado—conclude that between 3.6 and 7.9 percent of the methane released by fracing escapes into the atmosphere.[118] One study is based on "meta-studies" and life cycle analysis. It is the review of previous documentation verses current data, a study of studies. The latter attempts to examine the entire course of the gas, from extraction through end use.

A fourth study—actually, a response to the Cornell University study in 2010—challenges the assumptions and conclusions to these works. It suggests a further comparison to the life cycle for oil and its derivatives. This would include comparing the massively more efficient heat-conversion ability of natural gas and methane, over other energy sources. When examined with this issue in mind, the results look quite different.

Nevertheless, methane gas does exhaust from gas wellheads. It is far more dangerous to GGE buildup than CO2. Its presence in the atmosphere is eliminated in a few days, compared to CO2, which can last for 200 years. The use of multiple casings, with well-poured cement walls, reduces the estimated loss at the wellhead to virtually zero. Thus, properly designed and constructed casings can be the most effective tool in a drastic reduction in the release of the potentially dangerous gas.

Trucks are big, noisy, and smelly as they burn diesel. The partially decomposed hydrocarbon results in a black smoking effluent much despised by today's urban and rural dweller. A modern well pad may require an average of 5,000 truck journeys. As we have just observed, many firms are reducing truck cartage. The experimental use of compressed natural gas (CNG) trucks may be an effective "life cycle application" for the near future as the newest diesel engines are brought to the well bore. Truckers are happy to get six miles per gallon today, so the far better mileage standards of the newest CNG truck engines may be a quick-response economic tool.

Locals complain about the noise, sound, and traffic. Productive lands abutting roads have seen reduced yields. Roads are destroyed quickly under heavy loads. Reclamation and remediation of these roadways is, or should be, a part of every county's contract with unconventional play developers. This is a small price to pay for the extraction of this newfound wealth. As an aside, the average shopping mall requires 20,000 truck deliveries each year, every year. Where is the hue and cry to ban malls?

Health Effects—Workers and Local Citizens
Workers

Health, safety, and environment issues—HSE within the industry—are important to everyone involved. Large oil and gas companies, unbeknownst to the urban guerilla campaigner, spend millions on worker training, health insurance, and environmental awareness. Business owners and CFOs who are responsible for profits view these expenditures just as they view every other cost that must be managed. By far the best way to manage the health care cost for an employee roll is education/training. That is why Jake and his crew are justifiably proud of their 367 days of accident-free work. They take pride in being good and wise in their work efforts. Companies do the same. HSE is a significant budget item for most shale gas and oil firms, but HSE education spending goes a long way toward worker satisfaction and efficiency. It also helps the bottom line.

As we have seen, a frac rig is a dangerous place. These are big boys playing with big toys. The U.S. Department of Labor statistics for on-the-job injury in the profession shows a continued decline in reported serious injury, year after year.[119] For a small group in any profession, peer pressure can be strong. When your life depends on the actions of 15 to100 others in a challenging environment, working with very large machinery, you had better not screw up. When you do, it is immediately obvious. It has immediate consequences. This is not a desk job. This is real work in real time. You will be scorned for the slightest mistake. You will learn from the abuse of the team. It isn't easy out here, so the workers better get it right the first time, every time. These guys don't accept mistakes. Their lives depend on far better than "six sigma" performance action every day.

While visiting a frac rig, I watched as a new crewmember was putting drillpipe on the lift device for placement. It was to be the next piece in the drill string, connecting the bottom-hole assembly to the surface equipment. These are 45-foot lengths of steel, 11 5/8 inches in diameter, that are stacked horizontally in preparation for the connection at the tool joint to the previous pipe. The lift secures one drillpipe, raises it vertically, hoists it over the rotating drill string, and is poised there while the operator stops the rotation. As the string reaches deck level, he reverses it to clean the borehole, then slowly lowers the new drillpipe over the string until the roughnecks grasp it with tongs.

The new ground crewmember has put the drillpipe on the lift backwards. No harm is done because his fellow crewmembers catch it before the lift begins the next hoist. It does slow the whole team down about two minutes. That guy will never make this mistake again. He's young; he can take the heat. The discussion isn't pretty, but it works. You depend on the others. They depend on you. You learn or your get hurt. You learn or your mate dies.

Safety is the primary focus of the drill rig operator during a borehole drilling or a frac. A team takes real pride in "injury-free days." It's not about company policy or bonus pay; it's about your arm, your leg, your career, maybe your life. These crews take pride in their work, but also in their work ethic. Safety meetings are led not by honchos from the head office, but by crewmembers who have seen it, who have made the mistakes, who have lost fingers and friends. Most crews have seen serious accidents. You can watch some on YouTube, if you have a strong constitution.

Local communities have expressed concerns about health issues, particularly in communities that are close to or part of the urban landscape. Cancer, radiation, and water and air pollution have come up as important local concerns, perhaps enflamed by national NGO campaigns and media stories.

- Simon Lee, a professor of medical anthropology at the University of Texas Southwestern Medical Center in Dallas, hasn't found a spike in breast cancer rates in the north Texas area studied.
- The Pittsburgh Water and Sewer Authority has performed extensive tests for radiation level increases throughout western Pennsylvania, and found nothing beyond the normal range of background radiation.

- The U.S. EIA has demonstrated that coal-fired power stations are turning toward natural gas for generation.
- The IEA has also, as we have repeated, declared a significant drop in GGE within the U.S. since 2005 as a direct result of unconventional gas plays.[120]

Meeting the Challenges

Dave Collyer, President of the Canadian Association of Petroleum Producers (CAPP), outlined the public concerns surrounding hydraulic fracturing at a recent meeting. Producers must earn and maintain **"social license"** within the community. In terms of people, the health effects of hydraulic fracturing chemicals have been called into question, he noted. The surface footprint, induced seismicity, wildlife disruption, air quality during the extraction, processing, delivery, and end-use of hydrocarbon resources, and the potential for groundwater contamination through the migration of fracturing fluids: all must be considered when creating and communicating industry best practices.

According to Collyer, CAPP's guiding principles for hydraulic fracturing are as follows:

- Safeguard the quality and quantity of regional surface and groundwater resources, through sound wellbore construction practices, sourcing freshwater alternatives where appropriate, and recycling water for reuse as much as practical.
- Measure and disclose water use
- Support the development of fracturing fluid additives with the least environmental risks.
- Support the disclosure of fracturing fluid additives.

- Advance, collaborate on, and communicate technologies and best practices that reduce the potential environmental risks of hydraulic fracturing.

"We do firmly believe that shale gas is a game-changer," Collyer said at a recent Developing Unconventional Gas (DUG) Canada meeting in June 2012.[121]

These principles can be applied across the board, by the entire industry. They promote clarity and transparency. These can eliminate many of the false bogies in the media and from NGO slanted stories. Share the facts with the public and potential adversaries. As Lincoln said, "When I make my enemy my friend, I no longer have an enemy."

The IEA urges the following to move toward greater public awareness and acceptance of shale gas drilling. It quite openly calls these the **Rules of the Golden Age of Gas:**

- Greater disclosure
- Engagement with each local community
- Effective well monitoring systems
- Ongoing well, fracing, and spill-management designs
- Wise water management
- Elimination of methane emissions

These applications can enhance the industry's public image, placate (some) environmentalists, and should add less than 10 percent to the cost of each well. They will also, according to the IEA, significantly increase the development of global unconventional shale plays to the tune of more than a trillion dollars over the next two decades. This is real empowerment, real GGE reduction, real employment, real poverty reduction, and real health improvement.

Many try to solve the issues surrounding hydraulic fracturing. These are members of all of the communities involved: drillers, contractors, landsmen, leaseholders, regulators, and NGOs. Others either ignore or try to circumvent these important issues. Let's examine, at close range but anonymously, four entities from the frac world, and discuss their individual responses.

Our **first example** is a major land-leasing, publicly traded, company. It refused to honor lease contracts in several areas of East Texas and was sued by the leaseholders. The courts upheld the plaintive, yet the company refuses to budge, seeking to overturn the ruling in appeals court. The company hopes its legal budget is smaller than its lease budget—and that this same budget is vastly larger than that of the leaseholders. The legal battle has yet to reach conclusion.

Lessors are irked,[122] as people in the Utica Shale fields of Ohio are also facing similar "renegotiation" of their previously agreed-upon lease arrangements. Commentary from the firm is entirely wrapped in legalese, without respect for the landowners.

Given the serious nature of the energy discussion in America—and the hyperbolic commentary from many environmental and political quarters—this type of behavior reflects poorly on the industry as a whole. It is an example of the "evil corporation" doing its best to make a bad situation worse for small communities. This can set back all aspects of the energy conversation story. The new film, "Promised Land," limns this image of the unscrupulous corporate beast but goes further to imply that the evil shown is universal to all corporations. Should we ignore the fact that the government of Abu Dabi, a major oil producer in the Middle East, was a significant underwriter for the film? Had the film been in support of the gas industry, and this fact became known, what reaction could we imagine from the media?

We all have two ears and one mouth; let's use them in like proportion. Don't just hear, but really listen, and then speak. Respond to facts rather than fears. Witch hunts are so 17th century. Let's take an enlightened approach, examining the facts of each situation, listening to all involved, and attempting conciliatory understanding. The work in the fields of unconventional oil and gas exploration is a work in progress. Applying hard and fast rules to an evolving landscape stifles the evolution.

Quite a few bad eggs are in the energy discussion basket. Sad to say, there may be representatives from every community in this treasure trove of troublemakers. **As** our **second example,** a major environmental NGO has recently come out with their policy, "We should not drill for natural gas." Their oral comments before the EPA's Science Advisory Board, or SAB, in 2010 are quite descriptive. The NGO:

- Suggests the application of the **"precautionary principle"** when examining hydraulic fracturing operations and their potential links to groundwater contamination
- Suggests that **current regulations are insufficient** as applied by the individual states, whose pitiful, helpless bureaucracies are overpowered by little funding and small staffs
- Suggests that simply because "no causal links" can be determined, nevertheless, incidents of contamination should be approached with the **"basic tools of statistical public health analysis"**
- Urges that the EPA **extend its jurisdiction** to all areas of the nation and all laws not currently covered by previous statutes, such as the Safe Drinking Water Act.

While urging the application of "**baseline studies**," the NGO suggests these same studies be used to examine all regional differences in the hydraulic fracturing process.

The NGO describes the recent U.S. Department of Energy's text of April 2011, "Modern Shale Gas Development in the U.S., a Primer," as "misleadingly sunny," an "industry lobbying document," and states that the EPA should not "be relying upon industry public relations statements to inform." The publication is from government printing presses, directly from the DoE.

Which legal team would you like to take to the woodshed and beat with a shovel first? Both completely obfuscate and confuse the subject with legal attacks on word usage. The only person who can walk under a snake's belly is a lawyer. The absurdity of both positions makes you question the legal system:

- How can a multibillion dollar company try repeatedly to renege on formal legal contracts executed and agreed upon years before?
- How does a multi-million dollar non-profit seek political advantage from a legal dispute?

The final oral comments in 2010 are revealing: "Shale gas sits at the intersection of the climate change debate..."

Yes, it does. The IEA, the International Energy Agency has, as we observed throughout this book, recently stated quite clearly that shale gas development has been the major driver behind the U.S.'s 450-million ton reduction in GGE since 2005. By the end of 2012, the U.S. will be the only nation to have met the Kyoto treaty's demand for GGE reduction—despite the 98-0 refusal of the U.S. Senate to ratify the treaty.

Our **third candidate** for examination of content is another environmental NGO. In testimony before the same SAB noted above, they added their perspective in a 13-page discourse. Suggestions included expanding the current "study on hydraulic fracturing and water use" to be all-inclusive of the "life cycle process":

- Inclusion of storm-water runoff and erosion, produced water (deep water), wastewater treatment, "cumulative impacts," current industry best practices, seismic activity, threats to food supplies, reported events, and effects on landscapes, habitat, and community character
- EPA request additional funding from Congress and "take all action required to assure this study is unbiased, peer reviewed and free of political pressure from any special interest."

The remainder of the suggestions appears reasonable, as most are directly applicable to the water/frac issue under investigation.

The NGO's proposals may be taken as a reasonable approach to the issue. It certainly reflects their "legal habitat," and that should be expected. They do acknowledge the expected impact of natural gas development on the national economy and the "lower carbon future." While they stand firm against oil shale, tar sands, and liquid coal, they have no breach with "unconventional energy sources such as natural gas." They call for careful management of fracing byproducts. They frankly discuss the future of natural gas-powered energy sources for the nation, the current water risks, and the means to deal with these effectively.

Is it legally appropriate to represent both sides of a national debate? A recent article in the New York Times suggests it is not fair.[123]

Our fourth subject, an energy and environmental consulting firm, has a global presence and a long history of advising on environmental sustainability. Yet, because they have oil and gas companies on their client roster, their work is "tainted." The firm has worked closely with the EPA during both groups' formative years in the 1970s. The company grew with contracts to monitor the Alaska Pipeline, Love Canal, Superfund sites, and restoration of Gulf wetlands after Hurricane Katrina. Their federal contract work makes them overly dependent upon the Congressional budgetary process. The company grew or diminished in size and influence with federal contracts. As a result, they have expanded globally and across several industries, seeking work to secure sustainable development, while maintaining close ties to federal and state regulators.[124] It would be difficult to pick a better example of cross-cultural approaches to problem solving. Today, the firm has more than a thousand employees across the nation and around the globe working in an interdisciplinary approach to environmental issues, regulatory response, and sustainable development.

For this, they have been branded, called out, for their work within the oil and gas industry—despite the scope and range of their environmental work. It can be very difficult working in an industry with constantly changing borders, friends, and enemies. Such would appear to be the industry of some NGOs. Mitigating should take precedence over adjudicating. Discover solutions rather than deep pockets, please.

As previously discussed, searching for truths that simply support your point of view is **data mining**. Many consulting firms are employed across the spectrum of contract work. Because a firm has worked for a company whose activities are constantly under review bears little on that firm's decision-making ability. In fact, the HSE work done (and required) by every oil and gas company makes the case for further investigative integrity. Who knows the work better than those directly involved?

Casting personal or political aspersions upon anyone who disagrees with you is an **ad hominem** and personal/emotional attack. We need to go beyond this type of behavior. We need to be reasonable in our conversation, seeking information from many sources. Questioning someone's personal integrity because they disagree is schoolyard bullying, and nothing more.

Four examples of work in progress in the oil and gas industry. Bad eggs and good omelette's. These choices, which intended to illustrate the disruptions they cause and the support they encourage, are representative.

Political Challenges
Rationality usually prevails in Washington.
—A smiling lobbyist

This is where the conversation becomes difficult. The politics of any discussion is biased, by definition. So, hang on to your hats and try to keep your lunch down as we take a rollercoaster ride through the "amusement park" of the politics of oil and gas.

The politics and regulations applied to fracing are neither homogenous nor heterogeneous. They are a hodgepodge of noise. As this seems to be the way politics works, so be it. Every game has its rules. No one doubts the political power of the oil and gas industry. It warrants no crocodile tears. The industry makes a lot of money for a lot of people. It pays to play. Billions in taxes and fees are simply part of the rules. Billions in profits ensue.

Urban dwellers seem to have developed a squeamishness about the actions of the industry. NIMBY (not in my back yard) is an unfortunate addition to our vocabulary. Drill away in rural

Texas and North Dakota—no harm done there. Bring the rigs closer to town, and people start to notice. Interaction between the public, the local community, and industry should keep everyone on their toes. Locals do quite well, once the rigs are gone, the pad is clean, and the new roads and schools are in place. Landowners and shareholders receive their lease payments and dividend checks. Engineers, geologists, and toolpushers work the rock, learn from each hole, and improve the industry and the nation.

Fracing has opened old wells to new use. It has opened new reservoirs to exploitation impossible two decades ago. It uses significant amounts of water, but far smaller amounts of chemicals. Fracs have increased the percentage of shale gas supplies form one percent to 25 percent of the U.S.-sourced natural gas in just a few years. Fracing has fed new gaslines for chemical feedstock and power generation across the nation. Electricity supply is now primarily from natural gas. Trucks now run on it. Someday your car may do the same. The impact on the environment is substantial and significant, in a positive way. Local environmental hazards occur and must be handled; ideally, they should be anticipated and prevented.

Still, accelerating the phase-out of natural gas, nuclear power, coal, and industrial biomass, and driving a transition to efficient use of renewable, non-polluting resources is the main ambition of a group of public interest lobbyists.[125]

This aim strikes the author as extremist. The perspective has no bearing in the real world of emerging nations, energy production, and demand into the mid-21st Century. The nation, the planet, must look to every available and appropriate energy production source during this century. By mid-21st century, the entire human population may find itself at the levels of wealth, health and education that we currently enjoy. Policies and regulations should encourage all forms of energy production, while markets should price them according to each local demand and production situation. It is just as foolish to promote one energy source today as it was to do so for another energy source half a century ago. Governments are quite good at protecting national interests—and quite bad at picking winners in any business.

An example might be ethanol. According to *The End of Country*, a book critical of the gas industry, "between 263 and 2,100 gallons of water (are used) to create a single gallon of the biofuel, ethanol."[126] The gas from a Marcellus well produces between 2,000 and 17,000 more BTUs of energy than a gallon of ethanol.

Facts have a nasty habit of cropping up in unsightly fashion. How can we encourage such a wasteful fuel policy, both in generation and use, as biofuels? Billions of tax dollars each year are devoted to ethanol production. A substantial portion of the global cereals crop is now diverted to biofuels, raising the cost of production and consumption of food for everyone. This includes the very poorest in the world. These people are just emerging from starvation, yet we now threaten to plunge them back because of these misguided efforts.

The exploitation of crops for fuel takes food from the mouths of starving children; artificial, tax-driven demand for corn in the U.S. affects the market such that subsistence farmers are forced to sell their yields instead of feeding their families. This is not energy policy; it is bad science, technology run amuck, and a regulatory nightmare paid for by taxpayers who subsidize growers in the Midwest and across the globe to divert food to fuel. Growers are made handsomely wealthy and become morally compliant tools for greater exploitation. Stop the nonsense. Place the lands into food production, remove them from production entirely, or substitute another crop such as sugar cane (which at least results in greater fuel efficiencies), but please remove the regulatory glut of demand for an artificial fuel of negative value to mankind.

Natural gas was, until recently, a close ally of the movement away from a hydrocarbon-based energy production cycle. It was seen as the natural transition fuel to the hydrocarbon free energy world of the future. Green NGOs accepted tens of millions in tribute from the gas industry, only to leave it at the altar this past year. NGOs today have a range of positions, from "stop frac now" to "a new gas future." Those in the latter camp are reasonable.

Conversations between opposing viewpoints must begin with a few, ascertainable, common agreements:

- Where is common ground with folk who support such fuels?
- How does one communicate with those who simply refuse to discuss an energy source with clearly definable environmental benefits?
- If you ignore the obvious, you are left with—a fantasy.

We have the right to form and express our opinion in this great nation. We can listen to financial wizards like Madoff and Millken, industry leaders such as Sandusky or Steward, and politicians like Gore or Maddox. But if we simply believe something because we are told it is the truth, then we are the greater fool. We should form our own point of view, based on knowledge, facts, and problem solving by trial and error.

So, let's get beyond these dystopian always-negative views. Let's listen to facts, and encourage change where needed. Indeed, let's continue to grow with the challenges rather than stop any growth because of challenges.

These issues are best handled locally with wisdom and the long-lost art of listening. Their handling at the national level is less appealing. A cacophony of voices drowns one another out in opinion-strewn arguments. Facts play second fiddle to egos and agendas.

Tax Policy

Subsidies. The game is all about subsidies. As we've seen with biofuels, subsidies control—indeed, they have created—this market. Every player in the energy game gets some action from the nation's capital, from regional support, and from local taxing authorities. Never imagine energy to be a level playing field. It is strewn with the remains of politicians, companies and non-profits that never saw what hit them.

Both Standard Oil in 1896 and Solyndra in 2010 exemplify the "deer in the headlights" attitude of the current "big dog" of their day. Each was a dominant force in their field, each in its own way.[127] Each fell in a self-destructive fireball. The sin of pride is easily ignored, but it never fails to destroy.

Subsidies can create industries. They can maintain an industry participant much beyond its vibrant youth. If they are not withdrawn, if the newly born entity is not weaned, it will soon become a ward of the state. As such a ward, it can continue indefinitely. It will do so inefficiently. Why do Wall Street bankers, peanut farmers, rice farmers, chocolate producers and oil magnates need the subsidies they receive? They don't, quite simply. Each firm should be able to stand on its own—note the verb is not "allowed" to stand on its own two feet. Wall Street bankers and rice farmers have much in common: They are "rentiers," or rent seekers. In the South you call a dog that won't hunt an "egg sucker" because he would rather steal an egg from the chicken coop than hunt with the pack. These firms are egg-sucking dogs that won't hunt.

Politics allows the price of a gallon of gasoline to drive national energy decisions. The desired result defines the action. This crazy approach to energy may explain the patchwork quilt of laws, accounting standards, and practices each energy firm, each citizen, must endure in order to succeed. Sound bites drive policy.

The fact is, since 1960, the inflation adjusted price of gasoline has fluctuated between $2 and $4 a gallon. All else is noise.[128] The fact is that 600,000-plus U.S. workers are involved in the shale gas industry. Just one shale gas field deposit, the Barnett in Texas, has created 100,000 new jobs since the 1990s. The Marcellus supports 140,000 new jobs. Total tax revenues could rise to $3 trillion by 2030.[129]

Many in the energy industry are egg-sucking dogs. If you cannot design, build, maintain, and evolve your firm and its products or services in the face of taxation, competition, cost-cutting multinationals, or reverse engineering, tech-thieving Asian competitors, you shouldn't be in the game. Man up! Seeking the financial support of Big Brother should be an embarrassment, an abomination. It isn't. It's just another way of doing business. "Diplomacy is war by another means". Rent seeking has become business by another means.

Ah, politics. Politicians are often the conduit for these rentiers. Until rent seeking is illegal, we simply have to live with it. To do so, politicians have their role to play. They open the doors and grease the wells of, if not commerce, then inefficient transactions. Politics defines the manner in which much of the energy industry does its business, as it does in the world of banking. Congress drafts the laws; regulators implement them.

"Subsidies" are an age-old tool of the political class to extract money and power from industry, individuals, firms, and NGOs. "Energy subsidies" have been around since the late 1880s. In today's highly divided and testy Congress, the amounts are astronomical. According to the U.S. Department of Energy's recent report, "Direct Federal Financial Interventions and Subsidies in Energy in Fiscal year 2010,"[130] a variety of energy primary-source industries received subsidies. The $37 billion-plus in support included tax breaks, loans, loan guarantees, R&D support, home-heating assistance, and conservation programs. Trust funds, such as Black Lung, Super Fund Cleanup, Nuclear Waste Fund, pipeline and storage tank cleanup funds, and a dozen others are excluded.

Here is an abbreviated table from the report.

Coal	$1.4B	4%
Gas/NGLs	$2.8B	7.5%
Nuclear	$2.5B	6.7%
End-use	$8.2B	22.1%
"Renewables"	$14.7B	39.3%
Conservation	$6.6B	17.7%
"Smart grid"	$1B	2.7%
Total	**$37.2B**	**100%**

Alternative energy is the $22-billion-pound gorilla in the government "investment" world. Of the nation's renewable power generation sources, four percent receive 60 percent of the federal subsidies, and 84 percent of these go to wind. It appears the government has figured out how windbags are paid for their efforts.

Examine this 2012 chart of energy efficiencies, Figure 22, from the American Natural Gas Association.

Electricity Generation: Cost vs Evergy Efficency

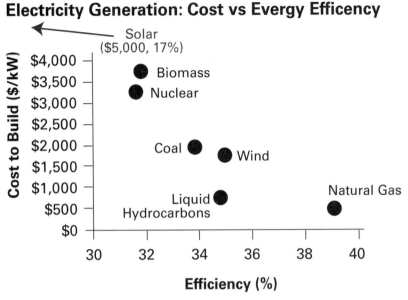

Note: Efficiency=BTU input vs BTU output

The arguments in favor of natural gas stack up very strongly indeed. The NGOs had a good thing going with their support from the gas industry. Both sides were beneficiaries of an ongoing policy debate that ultimately benefited society and the environment. This debate should continue. It is in the best interests of all parties to continue an active discussion, seeking fair and common ground, for the benefit of all. This behavior exemplifies the best in America, the industry, and the NGOs.

Lest the author be cited as pro-industry, the accounting of an energy firm (oil, gas, alternative, nuclear—choose your playground) is a wonderland of allowances, depreciation, depletion, tax credits, loan guarantees, and tax holidays. Read, if you can, the above-cited report on support for each of these industries: oil and gas, renewables, and conservation/smart grid. If ever there were a clear description of pigs wallowing in a tax trough, this report is it, in 3-D.

Taxes give and taxes take away. What both industry and politicians forget is that these taxes are taken from the citizens. Too bad citizens don't have more say in their use. Many refute the wisdom of a "flat tax," one tax rate applied universally, without exemptions, deductions, credits and the rest of the slop. Nations that have a flat tax—Hong Kong and Poland, for example—have greater economic freedom, enhanced tax revenues, and more business opportunity.

The U.S. Tax Code has thousands of pages of "tax items." Each of these tax items gives a business some slight or significant edge. Together, they allow "accounting standards" to be anything but standard. The Financial Accounting Standards Board (FASB) is a quasi-independent agency spawned from the SEC. It oversees the mathematics of business accounting standards. One must be adept at the highest levels of accounting behavior to enter here.

The Board is composed of chief financial officers from the very firms it oversees. It defines the acceptable activities of every accountant in the nation. It makes the rules. No harm here because someone has to do so. It is interesting that the chief foxes are in charge of the chicken coop. These regulators, in all good conscience, define, describe, and limit the uses of accounting. Under their watch, major advances have been made in the development of accounting. Under their watch, the likes of investor Bernie Madoff and Enron have stolen billions. Much remains to be done in this world of accounting.

Simplicity is replaced with subsidy. The last line from Shakespeare's "Romeo and Juliet" comes to mind. As the Prince of Padua comes upon the scene of lovers' carnage, he says:

"All are punished; all are punished."

The Tax Code

The phrase "tax simplification" has been bandied about for decades, to no obvious result. The Tax Code is like the mythological Hydra of Greek mythology that had a hundred heads. Cut off one, and another grew in its place. Change a tax law, and others replace it, with greater complexity. Only Hercules could destroy the regenerative Hydra, using fire to close the wounds of the severed heads before they could grow back. Absent a Hercules today, the Code persists.

The Code resembles the hydrocarbon deposits deep below our feet. The porosity is high, and the permeability is nil. Porosity refers to the rocks' storage capacity—it is the space between the rocks through which the hydrocarbons can flow. Permeability is their connectivity. How connected are the spaces between the rocks? The Tax Code has plenty of space between the regulations for any tax accountant or lawyer to work through. It has almost no connections within its complex structure. Recall the fabled Labyrinth: a maze wherein lay the Minotaur. So, too, the Tax Code, with its tens of thousands of pages, has no cogent, overlapping rational method—other than to tax.

As a result, tax agents are available to describe passageways. These may be members of Congress, lawyers, accountants, or even the IRS itself (its court system is notorious for creating or denying allowances). These agents have good intentions, and received excellent compensation for their efforts. The Code is serpentine enough to allow anyone access, for the right fee. Thus, another form of rent seeking permeates the culture. This monster is becoming a self-consuming beast. Corporations, NGOs, advisors of all renditions—each pays to play. The result allows many to pay no taxes—not just 47 percent of the voting public, but many large corporations. It forces others to pay an enormous amount of taxes, while granting certain tools for further tax reduction to others—oil and gas firms and alternative energy companies come to mind. Conversations become entangled webs of conceits: tax credits, government loan guarantees, depletion allowances, deductions, and exemptions. Much like the draft, it is equally unfair to all. The framers of the Constitution certainly had something else in mind.

Should we expect Hercules? Can we, as a body politic, change the beast ourselves? Time will tell. Looking at a copy of the first tax return, filed in 1917, it is simple. One page. It might be an interesting starting point for a discussion.

Science/Technology

Back in the desert, the rigs pull fuel from deep beneath the sands. This desert was once a seabed, rich with life. For tens of millions of years, this ocean floor accepted the accumulation of death from above. Dying sharks and shellfish, crustaceans and cephalopods added to the millennia of deposits on the ocean floor. As time passed, Mother Earth convulsed again and again. Seabeds became sandy shorelines, which erupted into hills and mountains, only to erode again. We can all see the convoluted results today when a highway is cut through the land. Layers of distinct joinery uplift and distort rocky ridges as we drive by.

The image many people have of an oilrig, as a straw dipping into a vast black pool and drawing oil via the derrick and pipeline, is a cartoon. The straw draws from below, yes. The pool is there, certainly. But it isn't quite as simplistic as the image suggests.

Gas is neither easily transported nor traded/substituted. It is a local commodity and is often "stranded." The only way to move it is via pipeline or ship. Pipelines are one-entry, one-exit transfer technologies; their receiving and delivery points are very specific.

Shipping requires liquefaction. This means tremendous capital expenditures for the liquefaction and regasification stations, and for the highly specialized ships. Qatar has spent $17 billion. It currently transports more than a quarter of the CNG globally. The Chinese and Japanese are the market, but China has its own shale and coal gas fields that it plans to exploit. Australia, Canada, Turkmenistan, Russia, the U.S., and African shale-rich nations are all planning LNG terminals, pipelines, or both. Over the next decade, global capacity may double, to 526 Tcf.

The technology evolves with the demand. While the heavy hand of governments in many countries suppresses shale-gas developments, others apply a looser rule. As an example of recent scientific application to a shale gas problem, Shell's newest vessel, "Prelude," is a floating LNG platform. It processes the gas, transports it and regasifies it, moving on when the fields are drained. It is the largest ship in the world, and is the precursor to an entirely new fleet.

The simple beauty of shale gas is it broad application. It is a true "flex-fuel." It can be used for energy production; heating homes, offices, and schools; and as feedstock for the manufacture of plastics and fertilizer. The development of the combined cycle gas turbine for power generation is the perfect example of applied science brought to bear. Its application to electricity generation has resulted in a far less expensive energy source, one that produces less GGEs. It is less costly to build than nuclear, and less time-consuming to permit and regulate. It is rapidly replacing coal as the primary energy source in the U.S. By the time you read these words, shale gas will have surpassed coal as the No. 1 source of generated power in the nation.

All of this has occurred with nominal government intervention in markets, the application of wise local permits and regulations, and the full participation of field workers, leaseholders, and communities. Irrespective of the EPA's desire to shut down coal, it may happen by attrition. Requirements for GGE-lowering technology seem like events from another planet. Shale gas leads the way to a carbon-reduced future through the wise application of science to the demands of oil and gas extraction technology—and most important, to the demands of the marketplace.

On the white-hot sand of Texas, each well pad stands alone, pulling oil or gas from the deep rock. Invisible lines beneath the desert floor connect the hydrocarbon to reservoirs, there to be sent to processing or burning for electricity generation. These tanks hold the immediate production for ultimate transshipment to the pipelines that lead to the refineries, where distilled petroleum is reborn as essential household, automotive, and industrial products .

These rigs may also send natural gas to the pipelines for delivery to electric utility plants, as more are being converted from coal and oil. The pipelines also send natural gas liquids (NGL) for processing and conversion to ethane, propane, and butane—the building blocks of agriculture and home heating. You stand anxiously by at the gas station, filling your tank, late for your next appointment, absentmindedly wondering whether humans are destroying Mother Earth. Ask the question differently.

How does this amazing Earth support a stabilizing population of creatures only recently and reluctantly freed from war and one which:

- Grows in wealth and health and intellect each generation
- Feeds itself healthfully with chemical additives from gas wellbores

- Transports its minions farther afield at ever-reducing cost with aviation gas, ship diesel, and automobile gasoline
- Adds years to each generation through chemical applications such as fertilizer

How does Mother Earth do it? With the intelligent assistance of her children, like Jake and his men.

ENERGY INDEPENDENCE

An investment in knowledge always pays the best interest.

—**Benjamin Franklin**

Three challenges stand out in the unconventional shale development saga as it unfolds: consequences, sustainability, and application.

What are the consequences of the development of national and global oil and gas fields? How sustainable are these consequences in the field? What further applications of the resultant energy production will present themselves in the near future?

Consequences Throughout the Global Economy

The current EIA statistics for daily U.S. oil imports show a decline of 40 percent from Middle East sources (Saudi Arabia, Iraq, Kuwait, and the Emirates) to 1.6M BOPD today. These should further decline to 806,000 BD by 2020, virtually disappearing thereafter. We are currently a net exporter of refined products such as diesel, kerosene, and propane. U.S. oil production, meanwhile, has increased each year since 2008 and is now at the level of 1998: 6M BOPD. The 40-year decline in U.S. production has reversed. Peak oil is an historical relic.

In his book *The Coming Prosperity*, Philip Auerswald suggests that the best is yet to come for America, for two reasons:

- The technological and organizational revolutions that have occurred in the U.S. are scattering across the globe.
- Their beneficial effects will provide similar but exponentially greater effects to the rest of the world.

Nowhere is this more evident or applicable than in the oil and gas industry. We sit at the top of the energy pyramid in terms of:

- Available resources
- Access to their development through legal ownership rights
- Technology and engineering expropriation techniques
- Skilled and experienced field and lab workers
- Distribution requirements to meet energy needs for the 21st Century
- Capital formation and deployment
- Deep and wide markets for product, energy, and capital
- An open architecture of communication
- Awareness and responsiveness to all aspects of the trade

We have addressed the first eight of these platforms in detail throughout this book. The last is more challenging, more elastic. Are we aware of and responsive to every aspect of the energy trade? Before we discuss awareness and responsiveness, let's briefly restate what we mean by "energy trade." Energy trade is the exchange of energy at a price. The price we pay to acquire a unit of energy fluctuates. It has done so in the past with changes in demand and in production, within the economic cycles of the nation. Today, shale gas at $3 per Mbtu and auto gas at $3.65 per gallon are the norm in the U.S. as of the summer of 2012. These prices will be different as you read these lines. The energy trade results from the myriad impacts on price each day. Let's examine these trades in detail.

The first trade in energy is in its expropriation. While this may seem a challenging word, expropriation is fitting. We are expropriating the resource to create energy for our species. *Electricity flows from shale gas wells.* How much simpler could it be? Electricity costs will decline as the availability and cost of shale gas increases. The unnatural economic consequence of this functionally infinite energy source—natural gas—is massive deflation. Deflation means a drop in price. This is not necessarily a bad thing. Your cost to drive a car today may be less than it was a year ago. Your cost to heat your home, school, or office building may drop. This frees up capital for you to spend or save, as you choose. Yes, other events are also driving up and down the price of energy, but they usually operate at the margins. [131] Their impact is far less consequential than the drop in the cost of energy production.

This is a primary cost. You must buy energy; you have little choice. When this primary cost declines, you discover "new wealth," or wealth you never had before. This creative capital acts as a nuclear generator. It costs you little. Its results and impact grow. If we see this cost decline continue, we will see more "new wealth." If we consume a similar amount of energy each year, we make more money from our passive behavior. This is discovered money. It flows into our pockets and into the economy of our family and our community. We use the metaphor of nuclear because its cost is low and its effects grow, more and more efficiently. We spend less and less each year doing the same thing. An economist would call this productivity gains. We have more spending money and implied savings.

Consumers are the first and best expropriators of energy costs. We are the end users. All along the supply chain, the cost reductions can be passed down while more are still being made at every link in the chain.

Thus the first "energy trade" is local. We trade fewer dollars each year for the same event. We make money at it. Its expropriation creates wealth at every level of society, and the consumers who pay less for a unit of energy today than they did last year are the winners. As the cost to produce power comes down, its retail price to end user, the consumers, also should decline. Consumers win.

The second energy trade is also local. It is the trade of worker experience for the hydrocarbon flowing from the wellbore. Drillers exchange their knowledge for the wellbore and its outpouring. Leaseholders capture a fee from the extraction process. Local communities initially capture a negative fee—a cost. Their roads, security, and peace of mind are initially disturbed by this noisy and dirty enterprise. The experience trade is the ultimate result. The drillers leave. The pad is "adjusted" back to its natural world. The leaseholders continue to receive their checks. Roads and communities are made whole. Water sources are improved, ideally left better than before. These results are neither spontaneous nor one off. They happen over years. The trade is more valuable each passing year.

The third energy trade is societal. As the newly added capital begins to function within the existing "capital society," this society expands. When people make more, they spend and save more. Taxes paid by the E&P, pipeline, and utility firms add to the tax base. Each community decides how to use this new resource. Similarly, increased wages and leaseholder payments are spent locally, adding to the capital society. Income for wage earners in Arkansas, Texas, North Dakota, Pennsylvania, and Ohio has gone up. Lifestyle follows. Look at the demographics for any county in the new shale gas fields, and compare these to their previous U.S. census figures from 1990. Much of the nation may have seen little or no wage increases (in fact, recent statistics show a reversion to the 1995 levels). These states are the exception. Put your cultural preconceived ideas aside for a week or so and pay a visit to northern Arkansas, southern Texas, the hills of northern Pennsylvania, or the plains of North Dakota. In place of "real estate for sale" signs, you will see one sign: Help Wanted. This may be Seamus McGraw's "End of Country," certainly. Very few people are complaining about the passing of poverty from their lives or their hometowns.

The fourth energy trade is national. An entire book could be written on this subject. The wealth creation process that kept Texas out of the recession is now beginning to affect the nation. The regional trade of labor, energy, and distribution affects much of the economy in each area. We have seen the massive inflows of capital and labor to the Bakken, Marcellus, Barnett, and the other fields. The significant increase in the wealth of the South is a direct result of plays in Oklahoma, Texas, and Louisiana. While much of the rest of the nation staggers up from the mat after the recession of 2008-09, these states and their workers are doing rather well. America earns new capital on her invested capital. Whether it is a dividend check or a new Chevy, it adds to the capital stock of the nation. Wealth breeds wealth.

The final energy trade is global. Capital moves across the globe at the speed of a keystroke. Long-term capital investment rises to the most current opportunity. As the perception of opportunity cost (risk) declines, capital moves to it. As water flows, so does capital. The biggest capital projects today are not dams or highways or spaceports or city redevelopment. The mega projects are in the pursuit of energy. Only the biggest players—the oil giants, the national oil firms, and some of the field support companies play at this game. Mozambique, Namibia, Angola, Brazil, China, and Kazakhstan are the playing fields. The plays are deep: 15,000 feet of water; then two miles of rock before the kickoff; then another six miles or more just to open the seam for the perf. A billion dollar rig pushes eight miles of pipe to open rock 250 million years old. This may take a decade. The winners expect results commensurate with the risks they take of time, capital, regulation, and politics. The next level down from these plays is regional global interests: ultra-large crude carriers and Suezmax tankers; pipelines across Turkmenistan to Turkey, across Ukraine to Europe; LNG terminals at several billion dollars each, servicing specialized ships at a cost of $200 million each. These projects take a decade to complete, and two decades to earn a profit. The workers number in the thousands, are usually local, and are trained professionals. Truck drivers, engineers, sailors, divers, heavy equipment operators, and roustabouts all support their families and their communities with wages earned far afield.

The transportation of hydrocarbons is, unusually, based on the sea. The mid-sized and smaller ships that carry the fungible fuel from the Middle East, Indonesia, and South America to the refineries of America, Asia, and Europe cost tens of millions to own and operate. Their profitable play brings jobs to seamen and traders— unsung heroes driving the cutting edge of technology, applied science, and wealth creation. Their ultimate power generation offers light, heat, and safety to entire continents.

For a fuller understanding and a fascinating account of oil and gas development at the global level, read David Yergin's books *The Quest* and *The Prize*. These trades across every level of life, from local to global, constitute the complex of the energy business. This is far more than a list of big firms with deep pockets. It is far more than simplifying regulations across all borders uniformly. It is far more than the mega or minor profits of hundreds of capitalists at work. Adam Smith's pin maker comes again to mind. It is not his intent to make the world a better place through the manufacture of his pins. The better place is a direct result of the complex of behavior that is his work. Even the most rigorous of regimes (North Korea and Venezuela) cannot completely destroy this innate human behavior. Such behavior is driven by little more than a worker's desire to improve the lot of himself and his family. Improvement is the common thread that ties us all into this intricate problem, the Gordian knot of energy production. Replace it if you dare; it will return, this cancer of wealth.

The Global-Trade Matrix

Awareness of and response to the global development of unconventional oil and gas will define the first half of this century. The intertwining of the trades just discussed, from home to globe, will create the next great lift in human success. The continued development of energy sources will be a primary key for a global population finally stabilized in numbers, yet growing still in habitat (the urban world will be the dominant human community by 2030) and in wealth. Wealth accumulation implies greater education, health, capital, political and legal demands by every human. All aspire to our situation. America remains the "shining city upon the hill." This aspiration will lead to an improvement in energy sourcing and its use. As personal economics migrates from survival to abundance, awareness of our family and our surroundings naturally occurs. Imagine seven billion conservationists/environmentalists with immediate access to virtually every political decision, offering constructive feedback and timely criticism across a array of cultural and personal demands.

The demands of such a population—our global grandchildren—will empower energy development. If 74 million Baby Boomers today are seeking dividends from good companies that are wise about shareholder and capital requirements, the demands of today's three billion youngsters must further the capitalist process. No command economy could possibly answer the call. Many have tried; each has failed.

This demand economy will drive further exploration and development. When every citizen of the globe has ownership in and receives income from energy production, all will benefit. When each company must answer to the shareholders of the world, they will all strive to meet the capital, energy, environmental, and local demands placed upon them.

This trade consequence can only lead to better exploitation technology, more efficient use of physical and human capital, and clearer results from the financials of the future. Admire both the future citizen and the future company. Go to the movies to see the fantasy evil corporate empire; pay for the ticket with a portion of your dividend check! Expect the energy company of 2050 to exert all possible environmental improvement as pro forma. Life cycle management will be as important as capital management, as pipeline flow management.

Mother Earth— or "Gaia"— is most capable of handling herself. The less time we spend on managing her affairs, and sticking to improving the lot of our own less fortunate, the better off she and they will be. Designing solar shades in space, and the casting of iron filings into the sea to manage her temperature, will soon be seen as the sophistic adventures of a young crowd of intellectual hooligans. Their games will pass, as they must, to the dustbin of history, along with their political origins: socialism, and communism. The market's invisible hand exerts more power. This invisible hand meets demands as they arise, almost as if anticipating them. The explanation? The simple mechanics of birds in flight, or of a school of fish. Entropy may be a rule in physics; in the history of humankind, it is the exception that proves the rule: achieve more; serve your family; aspire.

Sustainability

How sustainable are the massive increases in unconventional shale gas and oil? They will be in high demand. While we have perhaps sung from the hymnal, praising shale gas and oil exploitation as a paean to humanity's new potential, many concerns are clear and present. How can this ever increasing demand be met by an extraction process that is, by definition, depleting?

Visit an energy conference today. The room is full of rich men with expensive cars, pretty wives, vacation homes, yachts. Each thinks very highly of himself. They do not suffer fools lightly. The discussion of the dark side of initial production (IP) and decline rates (DR) is cloaked in jargon and hides behind obtuse graphs..

Look at Figure 23.

Haynesville Shale Type Curve

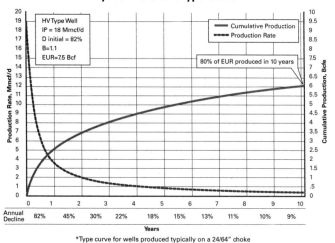

*Type curve for wells produced typically on a 24/64" choke

Let's compare the initial production, at the far left of the graph, to the production 12, 24, and 36 months after the play opens. An initial 18,000 Mcf/d drops to 3,500 by the end of the year, to 1,900 by year two and to 1,000 by year four. The log scale is distracting. If it were a normal scale, arithmetic-based, the drop would be devastating visually. By the beginning of the fifth year, 18,000 per day have become 850 per day. This particular well, and its counterparts that are illustrated here, are essentially worthless after five years. They will have to be re-fraced to source more hydrocarbons. This type of graph is common to the industry, and is found in virtually all public communications for publically traded firms. Nothing is hidden from view. There are no smoke or mirrors.

Several conclusions are clear:

- Many new wells have to be drilled each year to maintain an ever-increasing gas-and-revenue production level.
- The resource depletion rate far exceeds that of a conventional well; it can approach 45 percent a year.

- The rate of return (ROR) for the driller has to be very high simply to recover the massive capital infusion down bore. The reference figure is for $6 gas; today it is less than $4, which halves the ROR.
- The ROR figures for public consumption differ from the return-on-average capital figures employed (ROACE) by firms for financial consumption.
- The depletion implies the need to either constantly drill new wells, or redrill other zones higher or lower in the substrata from the same pad.
- The pipelines to these wells may have a limited usefulness if the "run is done" in five years.
- Fracing efficiencies must dramatically improve if the four to five percent recovery rates for U.S. reserves are to be fully exploited.
- The EUR assumptions are far off base today. Estimated ultimate recovery depends on capital, frac success, and decline rates.

Much of this conversation is arcane. Who really cares about the distinction between ROR and ROACE? As long as they keep paying dividends, most investors are happy with the flow—of gas and cash.

The literature in today's oil and gas world is often filled with debate about these decline, or depletion, rates, the contingent drive for more and more drilling, and this effect on natural gas prices.[132] At the time of this writing, the "rig count" in the field has moved. Very few rigs are drilling for gas, having moved to "wet gas" and oil. Bakken and Barnett are crowding out Marcellus and Woodford. Oil is far more profitable than gas, even as Bakken oil trades at a smaller discount to global prices.

These decline rates are not insurmountable issues technologically. There is a massive amount of gas beneath our feet. Its recovery will continue for several decades. The frac zones in each field (lateral stacking of hydrocarbon windows) are numerous. Recall that each frac penetrates no more than a few hundred cubic feet surrounding the bore. Jake's team could refrac a dozen times over 20 years through the same wellbore.

Nevertheless, the industry has overplayed its hand far too often in the past. The fear is that it is doing so again. The jury is out on the long-term sustainability of natural gas production. Time will tell whether recoverable reserves and EUR are actual figures or dreamscape. The gap between the agnostic Art Berman's 147Tcf figure for recoverable gas reserves in the U.S.[133] and EIA's estimate that is greater than 10 times this figure may be filled in by future drilling events.

The second sustainability concerns delivery. We have observed the need for 250,000 miles of pipelines over the next two decades. Until these miles are in place, bottlenecks will continue. If they become pervasive, they may impact pricing, curiously enough driving it higher during a glut simply because it cannot be effectively delivered. Firms such as **Delta Airlines** are acquiring their own refineries just to maintain a fuel source commensurate with their projected needs. The pipeline requirements are huge and will persist for a generation.

This infrastructure buildup has a positive element. Construction means jobs. Welders today earn more than $3,000 daily for their trade in the fields of pipe laying. Roustabouts, the young fellows working for Jake, can earn $60,000 for their first apprentice year in the field. Construction workers are building tank farms across the Midwest as fast as these field professionals can be found and hired. There is no shortage of work in the oil and gas industry. There is a shortage of workers: both unskilled and skilled.

The trades have much to offer the industry. Returning veterans have much to gain from local gas and oil fields pipelines. Think about it for a moment. In the military, you are taught the value of team effort, pride of work, the cost of small errors in judgment, the price of putting your foot down before looking. The dynamic vantage point is both individual and group. What better words to describe the efforts of Jake and his crew? Every commanding officer of each unit returning from Afghanistan should have a short list for recruiters in the industry. The skills developed for war are very effective for energy development.

Regulations are the third sustainability worry. States are quite good at working with the industry. They know the territory, the work, and the local demands. A federally imposed regulatory framework would standardize much of the regulatory work, making it far more difficult to apply locally. Another layer of regulations imposes cost, time, legal and personnel constraint. These expanded regulatory requirements can set back the industry while paradoxically reducing environmental benefits.

Sustainability of a national energy policy would be welcome. Clarity of purpose is today entirely lacking at the national level, except to encourage the "anything but hydrocarbons" approach to power generation. The worst example is perhaps the catfight between NGOs over the leasing of federal lands for energy development. The impact of the use of thousands of acres of desert in the Southwest for solar power generation has all the cats fighting over an invisible mouse. Some claim the terrain for creatures great and small. Others claim primacy to solar energy. Still others demand solar panels should be on the roofs of all the world before a single panel covers the desert floor. These clashes in court are a welcome respite from the usual legal drudgery of suspicious claims from out-of-state lawyers against wealthy energy plaintiffs. It is nearly impossible to make a case in favor of the successful company with cash in the bank, for it is better to settle for a few millions, and get on with it.

A policy of "all of the above" towards energy development—alternative, gas, oil, nuclear, and the future potentials of methane hydrates and fusion—would be a welcome, level playing field. Allowing the capital and energy markets to determine pricing for each energy source without significant regulatory interference would be a breath of fresh air. Removal of tax credits from all energy sourcing—credits, depletion, depreciation, front running capex, etc.—may allow production, job creation, and capital formation to expand. With this expansion would come primary additions to GDP and new tax revenues.

Uncertainty reduces the draw for capital. For example, if the same amount of investment can be made in two energy exploitation regimes, but one is fully supported by the regulatory regime, and the other is burdened with regulatory schemes, which would you choose? If the regulatory regime is local and supportive of the wise application of a time-honored procedure, the process can flow more readily. The process is first of capital, then of men and equipment, and ultimately of oil or gas for energy production.

Application

We have discussed at length the process of fracing. Let's examine one simple application of the end result, a plentiful supply of natural gas. As this supply increases, new uses for it will naturally develop within a capitalist culture. These are already appearing. Of the nation's energy use, 24 percent goes to transportation, in the form of gasoline, diesel, and aviation fuel. Imagine the substitution of methane for gasoline.

Many city buses already run on natural gas, and UPS and other long-haul firms are experimenting with natural gas engines to replace diesel. So it is not difficult to imagine that our entire fleet of buses and trucks might run on natural gas in the near future, for a significant savings. July 2012 prices show natural gas for trucks at $1.70 per gallon compared to $3.91/gallon for diesel.[134]

GTLs, or gas-to-liquids conversion plants, also offer potential for further energy independence and emissions reductions by creating LNG, liquefied natural gas. These plants, suggested for development in Louisiana, along the lines of the CNG plant in Qatar, may offer new choices. Certainly they offer employment opportunities in construction and maintenance. At an estimated cost of more than $10 billion, only "little countries" such as Shell can bear the capital lift.

The promising future of natural gas is evidenced by the formation of a new consortium. **Clean Energy Fuels** has partnered with long-haul truckers and service stations to create **America's Natural Gas Highway**.[135]**Shell** has recently announced plans to expand its "Flying J" facility and expand its offer of natural gas for fuel to truckers.[136] Upon examination, the simple elegance of this approach is apparent. Long-haul trucks tend to cover the same routes each week; their logistics makes this easy, and these repeating routes make their logistics simple. Place a natural gas refueling stop (also known as a gas station) every 250 miles. This is 70 percent of the typical range of a long-haul trucker. Equip the CNG trucker with the new hybrid or pure natural gas engines, with prices ranging between $35,000 and $70,000. Teach the drivers and station operators the safe delivery and use of the new fuel.

This "test road" runs today between the Port of Long Beach (PLB) and Las Vegas. PLB is the busiest port in the nation, and the County of Los Angeles welcomes the chance to both reduce GGE and the damage to their roadways. Today, 245 of the LNG stations and 71 percent of the CNG stations are located in California. The experiment is designed to the current road conditions. It would be hard to devise a more realistic test for the entire concept.

This experiment will go a long way toward proving the efficacy of CNG or LNG for overland transportation of goods. Because the trucking logistics across America are quite similar, the process should be replicable. Because the trucking community is so close-knit, the new trucks and their fuel will be either very quickly adopted—or just as quickly rejected. Again, only time will tell.

The EIA estimates a savings of 500,000 bbl of diesel for every 1 Tcf of natural gas replacement.[137] That goal may be a decade away; however, it could come sooner if these recent trials in California run true. The challenges are the cost of engine, fuel storage (both onboard and in ground), mechanical engineering and training, and national acceptability by the drivers, both independent and fleet.

Today's trucker can run for 1,000 miles between refueling stops, but only 250 miles between CNG/LNG stops. The addition of one additional fuel tank would increase this to 400 miles. The current cost differential is a significant draw to independent drivers, who are the majority of today's truck drivers. Fuel costs are a real concern. Reduce the refueling and the fuel cost, and the fleet may transition from diesel to gas far faster than the EIA estimates.

For further technological discussion on the truck fleet conversion, www.Climatetechwiki.org is an excellent source. **Westport** is the leading engine manufacturer along with **Cummings** and **Peterbilt. Waste Management** has been experimenting with the technology longer than any other fleet operator.[138] Their running commentary with drivers, fueling stations, and mechanics is invaluable. The challenges are expressed as real-time events, and the responses are direct from the field. This is an experiment quite literally "in field trials."

The true test of a market is the depth and breadth of its product lines. A visit to the open-air market in Kasgar, China, proves the point. The range of products and solutions for the local drover, herder, and agricultural Uighur is as varied as it was in Marco Polo's time. Every item needed, from rawhide saddles to hand-woven rope to camel hair hats to fresh food and clean knives are here. The herds are freshly driven to market, and the negotiations are serious and quick. Slaughter and consumption are spontaneous. From truck parts to camel carts, if you need it, you will find it. Haircut? The barber's chair awaits. Gown for the bride? It's just behind the rug merchant's stall.[139]

So, too, in today's truck market—online. The major manufacturers have adapted to the energy reality of California in the 21st Century. Every imaginable type of "green truck" is available. Conversions?[140] Available from dozens of independent entrepreneurs. Financing? Just down the page on the right. Shopping for federal tax credits? Want to build a CNG station? Driving a rig in need of replacement? It's all right there.

Now consider the natural gas car. This car exists today: the Honda Civic GX. It is a little pricier (by $5,000) than the normal Civic, and somewhat more expensive than the hybrid version ($3,600). For 2,000 test models, the 2013 Ram pickup truck will be flex fuel with gas and natural gas. GM promises to have test models of GMC Sierras and Silverados next year. The cost to convert is easily determined.[141] Want to find the local CNG station? Visit wwwcngnow.com.

The fuel is far less expensive than gas. The CNG at truck stops averages $1.40 per gallon less than gas. That is at least 35 percent cheaper in many high-tax states such as California, where the current CNG price is $1.84. In the Midwest, it is half the cost. These fuel savings can recover the upfront surcharge in about four years, assuming you spend $1,500 a year for fuel. Balance this against the current tank size and weight, which is larger and heavier. Application awaits future technology.[142] The adventurous can do today what drivers in Australia, Pakistan, Holland, and a dozen other nations are doing—drive on CNG.

The U.S. has fewer than 1,500 natural gas stations, and less than half are available to the public. This figure will grow substantially over the next four years. Many are private fleet stations for the likes of UPS, Waste Management, and the Houston International Airport. You can find stations on the Internet quite easily.

Several other barriers to market today may appear insurmountable. The scarcity of filling stations is primary. These cost $2 million to $3 million to build, and they have to be near a gas pipeline, or one has to be built to it. Of course, your electric car is rechargeable at home, so why not your natural gas car? It can be, with the Phill. This vacuum cleaner-sized in-home option simply needs 240-volt, 15-amp access, and the natural gas that may already be fueling your heating and cooking requirements. In six hours your tank is ready. Today the $4,000 unit cost, plus installation, is a turnoff. But, so was the electric-vehicle charging station of just a few years ago. Times change. Costs come down with production that's driven by demand. In Atlanta today, you can have one in your garage for just $60 a month, all thanks to the local utility, **Atlanta Gas Light Co.**

Some are trying to convert natural gas to a fuel that can be burned in today's engines. Pakistan and Iran do so now. Most of their vehicles are natural-gas powered. **Chevron, Royal Dutch Shell**, and others are considering building conversion plants, at price tags of $10 billion or more. The Silicon Valley entrepreneur firm, **Siluria Technologies**, is trying to mass produce a genetically altered virus that coats itself with metal, and then acts as a catalyst for the conversion of methane to ethanol. Your grandchildren may find these stations in a generation.[143]

Let's go back to our friends, the truckers, who are setting the pace for natural gas vehicles. Their real-time experiments may be the proving ground for our vehicles over the next 10 to 15 years. As natural gas supplies increase and prices decline, the steady expansion of truck stops, such as Love's Travel Service out of Oklahoma City, and Kwik Trip in La Crosse, Wisconsin, offer local drivers a real choice, and truckers a cost-effective alternative to diesel. You can join the CNG alternative, and impact your conservancy views far more effectively than with a hybrid or electric vehicle, so says the Environmental Car of the Year award for 2011 for the Honda GX.

CONCLUSIONS

Nature's music is never over;
her silences are pauses, not conclusions
—Mary Webb

This book has been written to increase your understanding of the complex process of energy development. Hydraulic fracturing and horizontal drilling have been the "set pieces" for illustration. The entire conduit is important to understand: the exploration, production, pipelining, fractionating, storage, transshipment, delivery, and ultimate use of unconventional shale oil and gas. The intent here has been to promote and provoke social awareness, and ultimately, understanding of wise hydraulic fracturing. This promotion and provocation can be achieved through education and consultation with all natural communities: workers, local citizens, leaseholders, public shareholders, regulators, politicians, NGOs, and the general public.

Social awareness means participation and involvement by every affected party—local communities, all involved industries, scientists, academicians, technicians, regulators, and concerned organizations. All interested members of society should participate in the conversation. No group should be excluded. You can no more exclude the technicians and scientists from the oil and gas industry than you can exclude NGOs concerns about the fracing process or academicians' work in their educational departments. All parties must sit at the table. Industry, community, academia, and regulators all pursue common cause, and all should be heard by the commons—you.

We demonstrate respect by engaging in conversation. We "dis" others by refusing to have such a dialogue. We can agree to disagree on any number of issues, but we must all agree to continue the conversation. Where can we find such dialogue today? Such groups as the **Breakthrough Institute, the Energy Institute, PwC, the Bipartisan Policy Center,** and others are leading the march toward greater understanding of U.S. energy exploration, production, and distribution, and the concomitant challenges and opportunities, by engaging in an ongoing, open dialogue with no forgone or ill-begotten conclusions.

Global Energy as the New Currency

The U.S. has neither exclusive control of the world's oil, gas, and NGL reserves nor exclusive domain over the technology to extract, transship, store, refine, and deliver hydrocarbon-based energy. We *are* among a very few nations that recognize the full subsurface private property rights of landowners. American law has led the way in many areas, and does so once again. Most nations abdicate mineral rights to the State, deny the basic concept of private property, or both. Many nations simply ignore legal niceties, and take what they wish. Virtually all (85 percent or more) of petroleum worldwide is owned by the State.

Much of Europe sits atop vast hydrocarbon reserves, but many states therein have banned the use of hydrofracing and horizontal drilling technologies. In the few places where exploration has been allowed to commence, regulatory delays and legal constraints continue to push past development. The continent's dependency on Russian gas, Middle East oil and its own coal deepens each year. A license to explore in Lancashire, issued in 2008 to a UK firm, may not see the legal light of day until 2014. This is simply for the license to look. France and Bulgaria have imposed an outright ban on hydraulic fracturing. This means that 640 Tcf of shale gas is off-limits to E&P.

This legal distinction serves our nation well. Clear title to the land and its mineral deposits affords America the global lead pole position in the race for unconventional energy. As the dollar is the global currency, our legal status affords us the top dog position for energy development. The new global currency is clearly expressed in the global capital markets. Capital flows to its demand. The demand is here. A global currency is emerging. Currency is a medium of exchange.

Apply the law to our subject of deep-rock hydrocarbons. If you own the land above the formation, you own the rights to the organic fluids and gases at the bottom of a wellbore. In most jurisdictions, you arrange to lease this right to a developer and receive a royalty, typically 12.5 percent of their gross revenue from the well. Now apply this to the areas of the country just beginning to be explored: Appalachia, the West Texas desert, the Badlands of North Dakota, the lowlands of Arkansas and Louisiana. These hardscrabble farmers, cattlemen, and ranchers have struggled for generations to make a comfortable life for their families and communities. During the past 20 years, and increasingly so during the past five, these people have begun to reap the harvest of the unknown wealth deep beneath their land. Movies like "Promised Land" begin with these laudable concepts and show them clearly.

Tax authorities in other nations may be playing the game of greater fool. The currency of taxes often goes unnoticed. Taxes and job creation in the nation are astonishing:

- All tax revenues from the hydrocarbon-processing complex exceed $85 billion in 2012.
- Employment figures nationwide in the oil and gas industry are currently 7 million.
- The oil and gas industry added more than 345,000 jobs nationally from 2007 to 2009
- It is expected to add another 85,000 in 2011.[144]
- As many as 1 million new jobs will be created in the industry
- As much as $3 trillion in taxes will be paid by 2035

This payroll incurs taxes at several levels: federal, state, local, Social Security, etc. These employees spend money, and pay taxes on their consumption. Taxes and fees today run $86 million daily to the federal government alone. Since 2000, $100 billion in federal taxes have been paid by the oil and gas industry. Who wins this game? The entire nation. Neglect it at your peril.

Recent Local Approaches

Local issues can and will be resolved through local discourse. State regulators are far better trained and experienced at drafting and ensuring compliance with directives that reflect the geology, topography, and communities of each locality than a federal bureaucracy sitting at distant desks. State officials and their laws have to be at least as restrictive as the EPA regarding well design, construction, operation, and maintenance for the EPA to stand aside and allow them to perform their oversight actions.

As an example, Bradford County, Pennsylvania, along with about a dozen other counties across the U.S., is establishing an association that will advocate for the safe and responsible development of natural gas.

A resolution passed by Bradford County commissioners stated that increasing the production of natural gas and oil in the U.S. will accelerate the economic recovery, and the safe and responsible development of natural gas and oil "is critical to powering our nation's future." Counties across the U.S. are excited about the association, said chairman of commissioners Doug McLinko, adding that it "will give counties where oil and gas is produced a national voice on oil or gas-related issues. Collectively we have a voice. Singularly, we don't." [145] Stakeholders from various sectors have responded to the Bradford County commission's invitation to join in the work.

The Oklahoma Corporations Commission attracted the attention of other states on issues of horizontal drilling, hydraulic fracturing, and groundwater protection. Landowners, environmentalists, mineral owners, various sectors within the oil and gas industry, the legislature, as well as state and local officials, have all played key roles in developing new programs that take into account the needs of all the stakeholders, while allowing Oklahoma's energy economy to grow.

Examples of this include new regulations that accommodate the drilling of longer laterals with horizontal drilling, resulting in a smaller, environmentally friendlier surface footprint—which also improves the cost-effectiveness of developing Oklahoma's shale resources. New rules were developed for standards and procedures to encourage the recycling of flowback water from drilling sites, reducing the demand on freshwater resources. [146]

A group made up of commission staff, rural water districts, municipal officials, counties, tribes, oil and gas operators, environmentalists, and others has worked to strengthen the protection of municipal water supplies while continuing to allow development of oil and natural gas resources.

"Clear, consistent and competitive" energy regulations in Oklahoma require two additional things: cooperation and collaboration. Oklahoma's place of prominence for energy investment is proof of the benefits of that approach.

Regulatory Impact of Energy Policy

Examples abound of the misuse, lack of use, or complete ignorance of regulations as they apply to any industry. At the same time, excessive application of regulations can lead to some serious nonsense.

Imagine a summer road trip with the family to see the sights of these magnificent United States. You leave Los Angeles seeking the far shores of Maine. Along your route you visit 28 states as you meander through the western desert, the Texas Panhandle, the Mississippi River plain, the Deep South, then turn north to explore the eastern littoral of the original Colonies. You complete your voyage of personal and historical discovery on the far reaches of Maine's rocky shore.

Did you think that the gas you put in your tank in California was the same as you purchased driving across country? At last count, 19 different types of gasoline are required by regulatory dictate, according to the state through which you are passing. Your engine will certainly be the worse for wear, as much from miles as from additives. Ethanol will destroy your valves and erode the crankcase. The definition of octane varies from state to state as to the additives blended into the gasoline. Your mileage will vary depending on these additive and their percentage of the total fuel consumed. Were you a fanatic about recording your engine's miles per gallon consumption, you would clearly note the differences. If you use the newly engineered oil, more strain yet impacts your engine. These "alternative" fuel types will only get worse (better?) as ethanol inclusion demands increase with time, from 15 to 20 percent in a few years.

Yes, the new cars today get better mileage because they use computers to monitor and regulate your gas consumption. More important, they are lighter. If you have a hybrid, your mileage increases compared to the "straight-up" gasoline engine. Your expense also increases. Have you had to replace that battery pack yet? Ever had a breakdown? Regulations command the market to produce and the consumer to acquire. The EPA hopes to eliminate the gasoline engine entirely.

The roads that you drive across are tax-enabled. Are they deep black asphalt or grey lusterless concrete? The latter, while more expensive, is far less reflective, or absorbent, of heat. If you are concerned about the planet's temperature, you may wish that concrete was the norm. It reduces the "urban heat island effect" (UHIE). Alas, you would be disappointed. Regulations require the black asphalt across much of America.

The price you pay for that simple tank of gasoline varies as you drive across that hot asphalt on a summer day. It varies with the chemical contents of the tank itself. As much as 40 percent of the price is a soviet collective of taxes. These typically add a dollar or more to the price of each gallon of your fuel. Out of gas at the border? Buy a gallon on the California side of the California/Arizona border, and you will pay $1.34 more than if you push your car across the border to Arizona. Buy 20 gallons, and you pay $26.80 less in Arizona. Do so for the typical 1,500 gallons of gas consumed over the year for the average driver, and you will have saved more than $2,000 a year.

You can decide whether the air is cleaner over California skies than Arizona skies. You may think that Kentucky roads are cooler than Missouri roads. You may wonder why your battery pack represents nearly 25 percent of the price and weight of your car. You may express outrage that the cost of grain for the poorest people on the planet is now beyond their reach, and starvation once again will consume their young children—all because government regulators insist that corn be used in place of hydrocarbons for fuel.

Regulations rarely affect their original cause. This is through no evil machination by socialist conspirators in charge of a dummy Congress. These laws were promulgated for the best of reasons, rational reasons. Regulators have our best interests in mind. We all know that the road our asphalt paves leads to hell. It was with the best of intentions that the Titanic had no watertight doors reaching the top of each compartment. The Edsel, Pinto, and Chevelle were designed with the best of intentions. Perhaps on your ride back to your home in California, you will pick up a few spare water tanks to carry you across the desert, much like my parents did in 1952 when they drove across country. Perhaps you will need to do so in the future as regulations impound water use in the desert.

Capital Configurations in Shale Gas

While prices in the natural gas market are in deep recession, liquids and oils remain strong. Acquisitions continue apace, as some leave and some enter, or re-enter plays, fields and basins. According to a new, private report from Merrill, "new supplies from natural gas and oil derived from unconventional shale production are now adding $1 billion daily to the U.S. economy." [147] Savings from reduced fuel consumption are matched by capital investment returns in the field.

There is an astounding amount of capital in the E&P world. Smart guys with 25 to 30 years in the field working through the drillbit, having had some success and monetized it, are plowing their new capital right back into the deep shale basins. They all knew these deposits were down there. They just needed George Mitchell to show them how to do it.

Multimillionaire E&P executives are the new capital barons of America. While young technology, media, and telecommunication boys (Facebook, Groupon, et. al.) are facing groping challenges in the real world of wealth, these riggers did it simple: man up. Walking the streets of Texas towns a few decades ago, each man in his own way discovered the joys of becoming a roughneck. Time passed. Fracing, 3-D seismics, and horizontal drilling are the new applications that came together at the exact right moment. Put technology to capital; then, add sweat and luck, and good regulators. Five years on, you have a very large table filled with very smart, very wealthy "old guys." Guess what: Old guys rule.

A well project can cost from $450,000 for a recompletion to $30 million or more for an offshore well, down 29,000 feet in a mile and a half of water. The typical shale play well in the United States today will run from $2 million in the Barnett to $6 million in the Bakken to $7 million in the Marcellus. It takes serious money to drill 10 to 40 wells each year, or more. Significant returns are expected, quickly, on this capital. Recovery rates can vary between nine months and two years. If you are earning less than 50 percent a year, you need to change the drill bit.

Similar rules apply across the oil and gas industry. Shipping, storage, fractionation, pipelines, compression, liquefaction, deep water drilling—all of these processes require massive capital in-

puts over significant time periods. Their time horizons tend to reduce daily political uncertainty to a lesser evil. Their capex budgets
are often larger than the GDP of most nations. The phrase "cap-
tains of industry" hardly captures the power of their boardrooms.
Being a visionary is a requirement, as is a thick skin and tough
personality. You want these guys at your back in a street fight, in
the deserts of central Asia—Iraq, Afghanistan—or on a rig. If you
ever wondered why serious people do not run for president, it is
because the game here at the corporate board level is so much more
challenging—and rewarding. Warren Buffet looks to them for direction. EPA administrator Lisa Jackson has nightmares about
them. Russia's Vladimir Putin wishes he had this kind of power.

This is a capital-intensive business, but it is also a personnel-
intensive business. The crew is tough, knowledgeable, and safety-
conscious. They pride themselves in the "lifting scene" as hard,
loud, and good. The toolpusher will know in a millisecond whether the new guy can pull a tong. This knowledge extends up the
wellbore to the "kelly" at the top of each firm. Most senior execs
earned their first paychecks in the field. Roughnecks, engineers,
and chem guys—each one has failed and succeeded on the rigs and
in the labs during their time in the field. Failure makes a man, like
the Texas heat. Talking about it means less than leaning into it.

As the frac world develops, well costs decline. New tools and
techniques drive productivity; from ceramics to seismics, cost
reductions impact overhead while increasing flow rates. Each firm
has a conscious attention to the details of production. Proprietary
proppants are designed to the play. Each company's HSE
department is often the testing lab for new ideas. Safety—his own,
and that of his mate and his crew—is every man's responsibility.
The oil and gas industry remains in the forefront of worker safety
improvements.

HSE requirements from most oil and gas companies are stringent and evolving. They are based on field experience and driven by field know-how. These men in the field are the point men in highest danger of injury, disability, and death. The evolution of health, environment, and safety issues occurs in real time.

Add to this the very obvious result to the bottom line of reduced accident rates, improved safety, and a sustainable commitment to an ever-improving environment. Yes, these are real profit centers when viewed from a balance sheet. Yes, capital expenditures committed annually to these arenas earn significant returns for shareholders. Dividends increase when health, safety, and the environment are accounted for wisely. Improving these improves the balance sheet and income statement.

The Tale of Two Georges

The oil and gas industry has done a brilliant job of staying ahead of the HSE curve internally, and most firms are to be commended. The industry and its members have done a poor job of communicating environmental remediation efforts externally—to the media, the public, and the increasing numbers of concerned members of urban society.

The industry must improve the storytelling; in fact, the story itself has to change. There is only a short time. If the public becomes emotionally involved in the perception of behavior, rather than in the facts, people, and work, the industry will have lost the race. If the experience of France or New York is repeated here, at the national level, all should hang their heads in shame. As **George Smith**, USGS geologist said a few years ago, "We need the moral support of the American people."

Who could have guessed the expansive development of these long-ignored yet crucial deep deposits beneath the shales? All owe **George Mitchell** a tip of the hat, an expression of deepest gratitude for his professional fortitude. Without his sheer drive, beyond the bounds of reason, past his own Board's expectations, below the Ordovician, these new pathways through the wellbit would never have tripped.

As the exploration and development of the shales of the Barnett, Haynesville, Fayetteville, Marcellus, and Bakken formations expand, each firm should heed George Smith's warning. Lesser mortals will challenge these rigs, these big men and their big machines, these deep reservoirs of power, these new capital barons. Our New American Heroes, the toolpushers, derrickmen, and engineers on today's pads, are just beginning to face their greatest challenge. It is not from below but from far afield.

Today's thought masons—urban guerrillas in the powerless towers of society—seek to discredit the work of these good men. Unknowingly ignorant, the intelligentsia assumes their sophistic answers to this 21st Century quest for energy production are the only, best choices. Their alternatives are anything but, these governmental dictats from distant towers.

The industry's task is to accept geologist Smith's challenge. It is always about the rock, first and last. These new American heroes, these lifting lads, need the moral support of every oil and gas executive. These young centurions provide the source of power for Everyman in our great nation. They do so at ever-reducing cost, with ever-increasing productivity, at constantly improving safety margins, with an extraordinarily positive effect on the environment.

Lack of knowledge on fracing creates misunderstanding. The emotional sale by many environmentalists often halts further discussion. Creating an enemy by false innuendo is the current "state-of-the-art" political procedure. Mobs rule by fear and loathing. It manufactures an enemy. Everyone "knows" that corporations cut corners, ignore the law, bribe and kill as necessary, all in pursuit of wealth. They must be controlled, for our own good, for the good of the nation. Offer protection, for a small sum. A small "tax" is affordable, yes? Of course, it does grow over time, but look at the protection you are getting for such a small sum; it's insurance against an "accident."

Distant lawmakers ignoring local geology and society regularly impose uniform regulations. Declaim against "oil and gas subsidies," while granting ever-increasing sums to wind, solar, and hedgehog energy production. Classic Chicago politics. Share-the-wealth stories are all the rage, these days. Tax those who have more than enough. The disparity of income has become a cancer. It must be stopped. Redistribution of income through putative taxation is the federal government's response. Tax and regulate. Take from those who have, and reapportion to those who need—and the devil take the hindmost.

The national debt has become a tool in this strategy of global redistribution. Increase it to pay for the needs of the entitled. We have an obligation. The massive accumulation by the new barons of wealth on Wall Street and in Texas is unjust. It is our new politics of equality.

This Hollywood version of the evil corporation has stormed the heights of intelligentsia. The frac beast must be brought under control, by regulation and by taxation. The energy industry had better find an active and positive response to these perceptions before they become another "new normal." There is no script that cannot be rewritten, no punch line that cannot be pulled, and no law that should not be reconsidered.

A View from the Bridge

Arthur Miller's play "A View from the Bridge" is the story of a close family torn apart by desire, resulting in betrayal and death. Let us hope this tragedy does not befall the family of American business, citizens, and interest groups. All will be betrayed, all will lose, and all will be punished.

Let me offer a different view. This is a view from a bridge between the various players on the energy stage: citizens, communities, industry, and government. Hopefully this metaphorical bridge offers a means of communication, a chorus overarching the stage.

After visiting frac rigs, pipelines, compressor stations, and C Suites, I have a few suggestions. These are meant to be both bold and respectful; the challenge is to alter the conversation according to the audience. The time is short. The industry has the capital, labor, and drive to succeed. An unworthy adversary that is seeking discord and salvation through regulation will impede it. Use these ideas as drillbits.

Industry Guideline Suggestions

1. IEA "Golden Rules" as common ground for discussion
   a. Full environmental transparency, measuring. and monitoring
   b. Engagement with each local community
   c. Careful drilling site choice to mitigate leaks
   d. Rigorous water-management assessments
   e. Target zero venting/flaring of gases
   f. Improve project planning and regulatory control

2. Water testing pre- and post-completion
   a. Establish habits of water testing

3. Transport management
    a. Reduced use
    b. CNG trucking—experiment, promulgate by example
    c. Road remediation by contract—as a goodwill gesture
    d. Pipeline use for water delivery and disposal

4. H2O alternatives and reductive technologies
    a. Recycling, reuse, state-of-the-art disposal functions
    b. Alternative proppant design, replacement of water with gas

5. Closer cooperation with state regulators
    a. Onsite inspection
    b. Enhancements to reflect best practices
    c. Open architecture whenever possible
    d. Voluntary disclosures

6. Direct involvement with NGOs that are supportive

7. Public communication
    a. Get beyond "too nice to bite"
    b. Education is good; continue to enhance education about the industry at the school level; grade school to grad school at the national level; educational ads sponsorships, scholarships

8. Involve veterans returning from the wars, for they:
    a. Understand the work and discipline
    b. Are active, involved community members
    c. Are loyal and patriotic
    d. Are educated and highly trainable
    e. Are politically invulnerable

9. Outreach to all citizens at every level
   a. Locals, NGOs, politicos, regulators, media
      –demonstrate communication and a willingness to recycle profits
   b. Set the table for the discussion by
      – defining where HSE is today,
      – how engineering is evolving with the technology, and
      – setting national industry standards

10. Visibility, openness, communication, and community sounding boards offering awareness vs. seeking approval

11. Tithing of net production wealth
    a. Willfully and voluntarily
    b. Start with E&P firms in Mozambique naturals for "paying their fair share"
    c. A percentage of net profits spent (not given) on infrastructure, e.g., 50% of tax load, or e.g., 10% of net profits
    d. Assume social responsibility for development
    e. Least-costly social laboratory
    f. Cheaper than government intrusion.
    g. Net giveback vs. government wasting of capital asset

No. 1 should lead the discussion. All others will follow in its path. These ideas are developed from and supported by the industry. They naturally support the further, global expansion of unconventional gas/oil plays. They are agreed to by several environmental NGOs.

The last is contentious, certainly. Yet, this is what a tax is, a "tithing" from excess capital back to the community. When it is forced, it is neither helpful nor useful. When it is voluntary, it is both efficient and measurable. Simply look at the economic health of the State of Utah. Tithing is a fundamental aspect of the Mormon Church. Every member gives 10 percent. The State of Utah constantly reduces it taxation while its citizenry does quite well, even during recessions. Why? Individuals and firms have been wire serviced to support where necessary. It simply isn't the business of the state.

Accept this lesson as a model for the oil and gas industry. Establish a willful sharing program in every community in which each firm is involved. Give back: road remediation, water disposal, training, and education. Do so beyond the scope of your firm's immediate impact. Do so willingly.

Many firms are doing so today. Follow their lead. The profits available from unconventional shale plays are enormous and growing. Give some back—or have more taken from you in the form of excessive taxation and regulatory obfuscation. Seize the day. Make an example of the industry for every other industry in America today. In 2009, the industry paid $10 billion in taxes, and from the same coffer, $11 billion was "spent" or "invested" in alternative energy. What does the nation have as a result? Gas at $4 a gallon, corn at $8 per bushel, and frac moratoriums. This nonsense must stop. Allow the invisible hand of the marketplace to determine the fair price for the demand of energy, and to determine the best sources for it. There is no set piece answer. All answers are allowed, if the hand of government will simply withdraw from unfair combat.

Consider these parting thoughts as we grow in this new energy revolution, this new energy Renaissance:

- Government should set the stage for the energy discussion, rather than determine who can speak.
- Industry should venture with its capital to the far shores of risk, in expectation of fair reward and just loss for extraordinary risk.
- Science and technology should continuously apply the most recent lessons from the field to each new drill site, in as open a manner as possible.
- Workers should ply their trade to the highest bidder in the roughest of professions.
- Regulators should hold firms to the highest standards set by each, agreed upon by both, while facilitating industry expansion.
- Communities, welcoming tax and wealth, should expect capital contributions to repair damaged public works and to infuse the local environment.
- NGOs should follow the lead of aware entities in the non-profit world: Work with the industry rather than shut out all discussion. Confrontation is so 1968.

Set an example. Follow a wise leader's urging of 2,000 years ago: "Render unto Caesar that which is Caesar's; render unto the Lord, that which is His." The more Caesar demands, the more he receives. Change the mantra. Set an example.

As we have observed, the EIA has stated that the U.S. has reduced its GGE by 450 million tons since 2005. The majority of this reduction is attributable to the gradual substitution of natural gas for coal for energy production. Even Barbara Arrindell's website, www.DamascusCitizensForSustainability.com, quotes several academic supporters who concur to at least a 25 percent reduction in GGE vs. coal use.[148] This fact alone should provide ample evidence of the "life cycle" importance of shale gas.

The phrase "sacrifice the good in support of the perfect" comes to mind.

When is good simply enough?

When does open discussion take precedent over harping polemics?

When does proprietary protection give way to open architecture?

The environmental groups have to acknowledge the importance and value of unconventional shale gas and oil development. The public must get beyond the evil empire image of corporations and corporate culture. The oil and gas industry has to demonstrate that their internal HSE evolution has everything to do with the concerns of NGOs and local citizens, and very little to do with "proprietary technology." Regulators must represent all citizenry, rather than the latest squeaky wheel.

Epilogue

The year is 2022. A decade ago, the nation was just entering the New Age. We had no idea what the future held. The sea of gas upon which we rest had just been touched with the first few thousand "fradrilles"—drills for fracturing. Little did we know then what we had done. Pandora's Box was open. Or, rather, the Gates of Heaven. We actually paid more than $3 per gallon for our fuel in 2012—10 times what the price is today. People spent enormous sums on "smart cars" that were anything but smart. There were lines of desperate jobseekers when a job was posted at the local coffee house. Most felt they were owed something by someone else, usually the government.

Today, we drive natural gas-powered vehicles of every shape and size. LNG is so prevalent and cheap that it is essentially free, like computers and electricity and education. Most of us have mini-lines connecting our garages to the gas main. We fuel up each evening. The gas station of the past is still there, but we use it only for long trips. Our 400-mile tank range (the new CAFE standard) is more than enough for most families' driving needs. If you had to make the comparison, we pay about 25 cents for the equivalent of a gallon of the old "gasoline."

Our high school kids have a choice. They can go to trade school and learn welding, pipeline maintenance, drilling operations, vehicle or mechanical design, or they can go to an engineering college to learn petroleum, chemical, mechanical, or aerospace engineering. They are "paid for grades," and the competition is fierce. Fourteen-year-olds learn tig and mig welding from 30-something senior leaders. Work is demanding and well-paid. Most enjoy and expect it. Those who'd rather work in the lab, at a desk, or in an office have choices aplenty. Finance, inventory support, and computer "gaming" that drives industry are all jobs in high demand. The choices for employment are wide and deep. Most find a change of career useful, and three careers during a 50-year work span are normal.

Plastics and chemicals are the new research areas for tomorrow. Adding gene pools to a hydrocarbon molecule is the newest video game—in 3-D, of course. The first cloned derricks have just been approved by the privately held FDA. The obvious combination of giraffe genes with high tensile steel has made lightweight, super strong yet flexible drilling towers. They say that the first submersible "living rigs" will be "3Ded" by 2030. Living refineries and plastics plants are more common today than you could have imagined. They are living only in the sense that their operations are driven at the Nano level by organic molecules with instructions imprinted by Nano-computers designing as the situation changes. A shame these weren't available for the oil spills of old: The "nannites" could have consumed the spill, held it for collection, and then used it to replenish a few tank farms. Yes, even these tanks are organo-molecular in design. They rise and fall with volumetric precision.

Cars are flex-icons. You can add more trunk space for a trip, or reduce it to a small urban vehicle in the time it takes to wash your hair in the shower. The program is stored in your home computer, of course. Color, seat shape, music, driving habit (of the auto driver), and audio preferences are programmed before your leave the "garage." For longer trips across the continent or overseas, you add a sleeping compartment and schedule a connection to the AirTran service. Your minibus drives to the aero port, ties in to the departing Tran, and is hoisted aloft by the sky service, all fueled by Nano carbon-recycled synthetic coal gas. They had tried the CO_2/H_2O matrix, but it produced too much rain.

Pipeline design is the challenge today. With nearly one million miles in place, the matrix has grown cumbersome to work around. The elections this year will determine which firm's "pipe dream" will be developed over the next decade. With instant voting and 96 percent participation, decisions are made quickly. Will the self-generating pipeline or the stasis fabric technology win? Your vote counts!

Energy production grows with the economy. Recession, booms, and busts come and go with no rational behavior. Complexity and Chaos Theory studies have been the rage in the post-doctoral world for years. Most students still refuse to accept that rational behavior is but one of many forms of economic behavior.

The real challenge for the future is the Space Elevator. The first multitrillion dollar enterprise, this Beast or Bounty may not begin rising above the plains of Golgotha for another generation. When it does, many expect a new age of enlightenment—while more than a few expect it to sink of its own weight into the core of the earth and destroy the planet.

The newer offshore rigs, resplendent with their mini-city landscapes, adorn our continental shelf. They are small harbors, hotels, restaurants, and shops for the shore-bound citizenry. RigWorld has become the popular dividend play in the market, earning seven percent for its shareholders, who appear to be nearly half the nation today. With the elimination of taxes on income from all sources replaced with the flat seven percent consumption tax and seven percent residence tax, no one files returns. The IRS will close in two years. Nothing to do. The departments of Labor, Energy, Education, HHS, and TSA are all closing. The challenge has been retraining nearly one million government workers in the principles of free enterprise.

Legal advice and lawyers have declined on the social ladder as their wealth thievery was finally exposed in 2019 during the Crisis of Confidence. The results were sobering for all. Now that laws have been reduced to local application, with instant acceptance via Internet juries of 12,000, the profession seems to be relegated to the historical junk pile of social scientists, climatologists, and talk show hosts. Some things just take a generation to go away.

Health care has improved, as well, since the early 21st Century. Robo docs and daily self-examination take the place of much previous "doctoring." Those who are seriously ill as a result of their past behavior now accept the consequences. If they can afford it, they have themselves taken care of. If not, they go to the terminal facilities and enjoy a few days of peace before their self-administered euthanasia. If their family or religion forbids this, they take care of their own. Responsibility for self is all the rage today. You wouldn't be caught dead asking for help.

The nation exports more refined fuel products today than ever before to rapidly growing "emerging nations" in Africa, Asia, and the newly reformed Near East. The capital flows to these nations is even greater than the fuel flows. Oddly, the newly minted wealth of the Southern Hemisphere supports more development there than the capital flows from America. The disaster that was Europa is slowly being resettled. The diaspora has finally ceased. The Dying Time of 2018-20 saw the virtual elimination of a generation of the populace, as they simultaneously succumbed to the idea that death was better for the planet. 100 million people took their lives over the course of 18 months. Historians will argue over the causes and consequences of such events: Climate fear? Genetic self-destructive urges? False beliefs in false gods? Or was it simply the Movie? Is it possible that a 3-D videogame available simultaneously to every iPhone on the planet could have so roused a people to self-destruction? The Curse of Allah. Curse indeed.

Humanity has nearly defeated absolute poverty. Starvation is a health choice today, never the result of political or economic foolishness. We all agree that a stable population goal of five billion is quite enough. Education is free, as is health care. The cost is so low that charging a fee for a Robo consult would be laughed at. Elders and those who enjoy teaching provide classroom and video classrooms for millions. Self-forming local groups evolve and dissolve as the need arises. Sometimes, these groups become small companies, cities, or cultural homes.

As for the New Cities...

Eat wisely. Sleep well. Love with abandon!

Acknowledgements

The unknowns behind a book are perhaps the most important elements in bringing an original manuscript drafted by an anxious author to you, the reader.

These people are wholly unknown to you, their work is often unrecognized. These magicians include copy, content, interior and cover editors, indexers and printers, wholesalers and distributors. Each fulfills a role in bringing a manuscript to life; each role is integral. The result is, ideally, a handcrafted work of artistry. If you find this book so, we shall thank the workers behind the scene. These include:

Dawn Klingensmith has edited the original manuscript superbly. Time and energy spent between his editor and her author is intense. It often seems that every word demands attention, every phrase, every footnote.

Bang Printing (bangprinting.com) has always done a fine job of putting these humble words to paper. They have earned their reputation as an outstanding printer for the small and mid-sized publishers of this nation. They continue to increase their efficiencies to produce better books at lower prices.

Gwyn Kennedy Snider has created both the interior design and the cover. GKScreative.com is an extraordinary site for professional design in layout and appearance. Your first glance at the cover often determines your decision to purchase a book. She has been successful if you now own this book.

Sharon, my dear friend and wife is the inspiration for much of my life, including this book. Her urging me on has allowed you to learn more. She saw what I could not: a vision of reaching out to others to help each reader understand. the book is a real gift from her to each of you, through me. Thank you.

Abbreviations Common in the Industry

Barrel(s) (42 gallons)	bbl
Barrels of oil per day	BOPD
Barrels per day	B/D
Million barrels of oil per day	MPD
Billion cubic feet per day	Bcf/D
Trillion cubic feet per day	Tcf/D
Barrel of oil equivalent	BOE
Estimated oil recovery	EOR
Exploration and production	E&P
Gulf of Mexico	GOM
Health, safety, and environment	HSE
Liquefied natural gas	LNG
Compressed natural gas	CNG
Two, three dimensional	2-D, 3-D
Blowout preventer	BOP
Natural gas liquids	NGLs
Environmental Protection Agency	EPA
Department of Transportation	DOT
Rate of return	ROR
Return on average capital employed	ROACE

A Few Definitions

Acid frac - A hydraulic fracturing treatment performed in carbonate formations to etch the open faces of induced fractures using a hydrochloric acid treatment.

Aeolotropic - the space between the drill bore, casing, or tubing

Annulus, annular - the space between the drill bore, casing, or tubing

Aquifer - any water-bearing rock formation, and special care has to be taken to eliminate any contamination by drillers

Backflow - fluid flow in response to pressure differential down hole

Biocide - an additive to kill the bacteria that could impede fluid flow

Blowout - an uncontrollable flow of fluids into the wellbore or, catastrophically, to the surface

Borehole - the wellbore itself, including any uncased section

Bottomhole - the base of the wellbore, at the edge of the drill bit or the end of the drill string

Butane - one of the NGLs found in drilling for hydrocarbons, along with methane, propane, and ethane

Caprock - dense rock overlaying another deposit

Casing - steel pipe cemented in place once the drilling has been completed

Christmas tree - set of valves, fittings, chokes, and pressure gauges fitted to the wellhead upon completion

Cofferdam - containment structure to hold drilling fluids and protect the surrounding area from them

Coiled tubing - a continuous length of pipe, up to 15,000 feet long, that is wound on a spool and fed directly into the wellbore; replaces single section pipe

Depletion - the drop in hydrocarbon reserves from production; can be contained by water injection; tax code definition: reduction in cost taken against income in first years of production

Derrick - the vertical structure in support of the drill string; can be self-contained or single-purpose

Drill string - the combination of drill pipe, bottom hole assembly, and any other devices employed in turning the drill bit

Field, play, zone - area of hydrocarbon

Fish, go fishing - to recover lost or broken parts

Flowback - flow of fluids from the well under pressure

Frac - the hydraulic fracturing of rock surrounding the completed wellbore at specific depths using specific tools

Guar - seed of the guar plant used in frac liquids to intensify sand-carrying capabilities; one of several proppants

Hydrates - methane trapped in ice crystals for which no extraction processes are currently known; when they are developed, the methane released could be a new energy source for the future

Hydrocarbon - a naturally occurring organic compound of carbon and hydrogen; gas, oil, and coal are the most common forms

Kelly - long steel bar transmitting rotary motion to the drill string

Kickoff - changing the direction of the drill string to the horizontal; the tool to do so

Life cycle - attempting to observe and account for all costs in an energy project; typically inclusive of supposed or expected economic and environmental impacts

Methane - one of the BGLs resulting from a well

Midstream - the production, processing, and transshipment of wellbore liquids from the top of the bore to storage facilities or end users

Monkey board - small platform near the midsection of the derrick upon which the derrickman stands when tripping pipe

Mud - drilling fluid used to lubricate the drilling operation downhole

Mud motor - directional drilling motor that drives the drill bit during horizontal drilling

Mud pit - tank that holds drilling fluids extracted from well; few open pits are in use today; most are steel tanks

Natural gas - methane

NORM - Normally Occurring Radioactive Material; its presence is common throughout the mantle and crust of the earth

Operator - drilling project manager; site manager

Perforate (perf) - design and execution of holes in the well casing to permit hydrocarbon flow

Permeability - ability of complex hydrocarbon molecules to move through rock; major factor in downhole fluid flow

Pig - rubber or plastic pipe insert used to monitor and service pipelines; smart pigs are telemetric monitors

Pipeline - tube used for transportation of fluids from wellbore to next stage in processing

Porosity - amount of void space in deep rock

Proppant - particles deposited downhole to hold open fractures after a frac treatment; a contraction of the words "propping" and "agent"

Recompletion - re-drilling of a previous wellbore for further extraction

Reservoir - deep underground naturally occurring storage area for hydrocarbons

Rig - equipment used to drill a well

Rock - affectionate term for the deep rock in each well

Roustabout - unskilled manual laborer on well site

Seismic - waves of 1-100 Hz used to study composition of the deep rock

Shale - sedimentary rock in which gas and oil is found

Slick water - saline slurry enriched with proppants used for fracturing

Sphericity - pertaining to roundness

Steerable motor - directional drill bit controlled from above

Toolpusher - rig boss

Tongs - massive self-locking wrenches used to hold the drill string, additional drill pipe

Trip - pulling drill string up out of the wellbore, usually to replace a worn drill bit

Upstream - hydrocarbon exploration and extraction

Water injection well - wells, usually old, depleted ones, for holding used water from a drilling operation

Wellbore - the entire drilled hole

Wireline - electrical cable used to lower tools and instruments downhole

Reference Sources

Bipartisan Policy Center

Founded in 2007 by former Senate Majority Leaders Howard Baker, Tom Daschle, Bob Dole, and George Mitchell, the Bipartisan Policy Center (BPC) is a non-profit organization that drives principled solutions through rigorous analysis, reasoned negotiation, and respectful dialogue. With projects in multiple issue areas, BPC combines politically balanced policymaking with strong, proactive advocacy and outreach.

Websites

Oil and Gas Association Links
- American Association of Petroleum Geologists
- American Association of Professional Landmen
- American Gas Association
- American Institute of Chemical Engineers
- American Petroleum Institute
- Association of American State Geologists
- Geological Society of America
- Independent Petroleum Association of America
- Integrated Petroleum Environmental Consortium
- International Association of Drilling Contractors
- Mineral Information Institute
- National Association of Division Order Analysts
- National Association of Energy Service Companies
- National Association of Royalty Owners
- National Technology Transfer Center
- Stripper Well Consortium
- Texas Alliance of Energy Producers

- Oklahoma Independent Petroleum Association
- Oklahoma Mid-Continent Oil and Gas Association
- Organization of Petroleum Exporting Countries
- Panhandle Producers and Royalty Owners Association
- Permian Basin Landmens Association
- Petroleum Technology Transfer Council
- National Energy Education Development Project
- Rocky Mountain Mineral Law Foundation
- Society of Exploration Geophysicists
- Society of Independent Professional Earth Scientists
- Society of Petroleum Engineers
- SPE-Online Information Library
- Society of Petrophysicists and Professional Well Log Analysts
- Society of Women Engineers

Federal Government Links
- United States Bureau of Land Management
- United States Department of Energy
- United States Department of Energy National Energy Technology Laboratory
- United States Department of Energy/Energy Information Administration
- United States Environmental Protection Agency
- United States Geological Survey
- Environmental Compliance Assistance System
- Minerals Management Service

References
Petroleum Related Rock Mechanics, Bernt Aadnoy: Gulf Professional Publishing, 6/2011

Oil and Gas Production in Non-Technical Language: Martin Raymond and William Leffler: 2006

Oil and Gas Pipelines in Non-technical Language: Thomas Miesner and William Leffler; Pennwell, 2006

Nontechnical Guide to Petroleum Geology, Exporation, Drilling and Production: Norman Hyne, Penwell, 2001

Oil 101: Morgan Downey; Wooden Table Press, 2009

The Deep Hot Biosphere, Thomas Gold; Copernicus Books, 2001

The Quest, Daniel Yergin, Penguin Press, 2011

The Prize: Daniel Yergin, Penquin Press, 1991

The End of Country: Seamus McGraw; Random House, 2012

Under the Surface; Fracking, Fortunes and the Fate of the Marcellus Shale: Tom Wilber, Cornell Press, 2012

The Coming Prosperity Philip Auerswald: Oxford University Press, 2012

Complexity: Melanie Mitchell, Oxford University Press, 2009

Understanding Complexity: Scott Page, University of Michigan, 2010

Careers

A wide variety of job search functions exist across the multidisciplinary world of energy. Here are just a few. Good hunting.

http://www.Pennenergyjobs.com
http://www.rigzone.com/jobs/search_jobs.asp
http://www.downstreamtoday.com/jobs/
http://www.oilcareers.com/worldwide/
http://oilandgaspeople.com/
http://www.simplyhired.com/a/jobs/list/q-Oil+And+Gas

http://www.pioga.org/careers/ (PA)
http://www.jobs.state.ak.us/energy.htm (Alaska)
http://houston.jobing.com/ (Houston, TX)

There are dozens of other contacts. Look at individual state employment sites, company sites or publication sites. This information is provided without charge or compensation. Good luck!

Journals

Hart Energy and Pennwell are the main publishers for the industry.

Hart Energy publishes:

Oil and Gas Investor Journal, E&P, Midstream Pipeline and Gas, Fuel, a wide number of directories and custom pieces for the industry and sponsors the DUG conferences.

Pennwell publishes:

Oil and Gas Journal, Oil and Gas Financial Journal, Offshore, as well as a wide variety of journals and training for the oil and gas and other industries.

Worldwidedrillingresource.com is another important journal.

Rigzone has an extensive catalog of current job offerings worldwide as well as online training in the basics of oil and gas.

Films

Gasland I, II
Switch—http://www.switchenergyproject.com/aboutfilm.php
Spoiled—http://www.spoiledthemovie.com/
Truthland http://www.truthlandmovie.com/watch-movie/

Appendix One
MLP Securities List

Name	Exchange: Symbol
NATURAL RESOURCES, OIL AND GAS:	
Midstream Operations, Compressing, Refining	
Access (formerly Chesapeake) Midstream Partners, L.P.	NYSE:ACMP (formerly CHKM)
American Midstream Partners, LP	NYSE:AMID
Atlas Pipeline Partners, L.P.	NYSE:APL
Blueknight Energy Partners, L.P.	NASDAQ:BKEP
Boardwalk Pipeline Partners, LP	NYSE:BWP
Buckeye Partners, L.P.	NYSE:BPL
Calumet Specialty Products Partners, L.P.	NASDAQ:CLMT
Central Energy Partners, L.P.	OTC:ENGY.PK
Cheniere Energy Partners	NYSE AMEX:CQP
Compressco Partners, L.P.	NASDAQ:GSJK
Copano Energy, L.L.C.	NASDAQ:CPNO
Crestwood Midstream Partners LP	NYSE:CMLP
Crosstex Energy, L.P.	NASDAQ:XTEX
DCP Midstream Partners, LP	NYSE:DPM
Eagle Rock Energy Partners, L.P.	NASDAQ:EROC
El Paso Pipeline Partners, L.P.	NYSE:EPB
Enbridge Energy Partners, L.P.	NYSE:EEP
Energy Transfer Partners, L.P.	NYSE:ETP
Energy Transfer Equity, L.P.	NYSE:ETE
Enterprise Products Partners L.P.	NYSE:EPD
EQT Midstream Partners, LP	NYSE:EQM
Exterran Partners, L.P.	NASDAQ:EXLP
Genesis Energy, L.P.	NYSE:GEL
Holly Energy Partners, L.P.	NYSE:HEP

Inergy Midstream, L.P.	NYSE:NRGM
Kinder Morgan Energy Partners, L.P.	NYSE:KMP
Magellan Midstream Partners, L.P.	NYSE:MMP
MarkWest Energy Partners, L.P.	NYSE:MWE
Niska Gas Storage Partners, L.P.	NYSE:NKA
Northern Tier Energy LP	NYSE:NTI
NuStar Energy L.P.	NYSE:NS
NuStar GP Holdings, LLC	NYSE:NSH
Oiltanking Partners, L.P.	NYSE:OILT
ONEOK Partners, L.P.	NYSE:OKS
PAA Natural Gas Storage, L.P.	NYSE:PNG
PetroLogistics LP	NYSE:PDH
Plains All American Pipeline, L.P.	NYSE:PAA
Regency Energy Partners LP	NYSE:RGP
Rose Rock Midstream, L.P.	NYSE:RRMS
Spectra Energy Partners, LP	NYSE:SEP
Sunoco Logistics Partners L.P.	NYSE:SXL
Targa Resources Partners LP	NASDAQ:NGLS
TC PipeLines, LP	NYSE:TCP
Tesoro Logistics LP	NYSE:TLLP
Transmontaigne Partners L.P.	NYSE:TLP
Western Gas Partners, LP	NYSE:WES
Williams Partners L.P.	NYSE:WPZ

NATURAL RESOURCES, OIL AND GAS: Exploration & Production

Atlas Energy, L.P.	NYSE:ATLS
Atlas Resource Partners, L.P.	NYSE:ARP
BreitBurn Energy Partners L.P.	NASDAQ:BBEP
Constellation Energy Partners LLC	NYSE:CEP
Dorchester Minerals, L.P.	NASDAQ:DMLP

EV Energy Partners, L.P.	NASDAQ:EVEP
Legacy Reserves LP	NASDAQ:LGCY
Linn Energy, LLC	NASDAQ:LINE
LRR Energy, L.P.	NYSE: LRE
Memorial Production Partners LP	NASDAQ:MEMP
Mid-Con Energy Partners, LP	NASDAQ:MCEP
Pioneer Southwest Energy Partners, L.P.	NYSE:PSE
QR Energy, LP	NYSE:QRE
Vanguard Natural Resources, LLC	NYSE:VNR

NATURAL RESOURCES, OIL AND GAS: Propane and Refined Fuel Distribution

AmeriGas Partners, L.P.	NYSE:APU
Ferrellgas Partners, L.P.	NYSE:FGP
Global Partners LP	NYSE:GLP
Inergy, L.P.	NASDAQ:NRGY
NGL Energy Partners LP	NYSE:NGL
Star Gas Partners, L.P.	NYSE:SGU
Suburban Propane Partners, L.P.	NYSE:SPH

NATURAL RESOURCES, OIL AND GAS: Marine Transportation

Capital Product Partners L.P.	NASDAQ:CPLP
Golar LNG Partners LP1	NASDAQ:GMLP
Martin Midstream Partners L.P.	NASDAQ:MMLP
Navios Maritime Partners L.P. [1]	NYSE :NMM
Teekay LNG Partners L.P.	NYSE:TGP
Teekay Offshore Partners L.P.[1]	NYSE:TOO

NATURAL RESOURCES, COAL, OTHER MINERALS, AND TIMBER

Alliance Resource Partners, L.P.	NASDAQ:ARLP
Alliance Holdings GP, L.P.	NASDAQ:AHGP

CVR Partners, LP	NYSE:UAN
Hi-Crush Partners LP	NYSE:HCLP
Natural Resource Partners L.P.	NYSE:NRP, NSP
Oxford Resource Partners LP	NYSE:OXF
Penn Virginia Resource Partners, L.P.	NYSE:PVR
Rentech Nitrogen Partners, L.P.	NYSE RNF
Rhino Resource Partners LP	NYSE:RNO
Pope Resources	NASDAQ:POPE
Terra Nitrogen Company, L.P.	NYSE:TNH

Appendix Two

Thomas Gold disputes the cartoon characterizations of hydrocarbon sourcing. He posits that hydrocarbons have little if anything to do with rotting plants and dinosaur carcasses. The hydrocarbons of methane, petroleum, ethane, propane, and butane are primordial constituents of Mother Earth. They are the soul of Gaia. His book, *The Deep Hot Biosphere* is a page burner if you enjoy chemistry, physics, astronomy, and cosmology. He is a multiple Nobel Prize Laureate, so dismiss his ideas at your own discretion.

His position is as follows. The planet was formed from debris surrounding the protosun that was to become Sol. It coalesced intermittently over a billion or more years. As it did so, it cooled differentially, rather than simultaneously. He demonstrates that all stellar bodies—planets, asteroids, moons, even the stars themselves—contain quite similar amounts of hydrocarbons. These are the basic chemical building blocks of the universe. As the planet accreted, it began to rotate and gravitational forces pulled more debris into its orbit. Over time, it developed a spheroid shape and effervesced through volcanic action and tectonics.

The predominant chemical, methane, trapped throughout the coalescing debris began a perpetual cycle of transfusion through rock pores to the surface. Upon reaching this molten layer, it was released and quickly interacted with oxygen to form carbon dioxide and water. Trace elements, as well as fundamentals such as helium, followed the rock pathways and are today still found only in concentration with the parenting hydrocarbons. Their presence, and that of a uniformly diffuse amount of two forms of carbon (C12 and C13) he offers up as solid evidence for the process he calls abiogenic diffusion.

He suggests that life itself was originally based on methane rather than oxygen, hydrocarbon rather than carbohydrate. The vast extent of life at the deepest and hottest spots on the seafloor is evidence for the Archaea life form, an entirely new branch of life. These thermomethanotrophs reproduce in an environment entirely hostile to life as we know it, yet they thrive. There is no genetic connectivity to any other life form. They are the original life forms, from which we have evolved.

As the Earth matured, oxygen and carbon dioxide became sufficiently prevalent in the atmosphere to encourage further life development: us. Meanwhile, these precursors to life lived on the methane extrusions from deep within the Mother Ship. This ef-flusion continues today. It is the source of all petroleum, natural gas, and NGL deposits. Its constant upwelling is the only possible explanation for at least seven characteristics of the energy fields throughout the planetary surface.

1. The fields are found in long geological patterns across differential subsurface structures.

2. The fields replenish themselves; despite more than a century of drilling and draining and discovery, there is far more oil and gas than ever imagined—and much of it is in decades-old wells that should have depleted long ago.

3. Hydrocarbon-rich areas tend to richness across all depths irrespective of geological epochs. There is simply too much of it too deep to be accounted for by rotting carcasses.

4. Hydrocarbons are found in non-depository fields of igneous and basalt concentrations, as well as across the deep ocean where no sedimentary events could have occurred.

5. Complex chemical signatures are universal rather than gradient according to depository action.

6. This also validates the universal presence of C13: Were it deposited by glandular surface action, the concentration should change with the depth signature.
7. The universal presence of the inert gas helium with all hydrocarbon deposits can only be explained as a vertical migratory event polished by the trail of forerunning methane.

His arguments are theoretically pervasive. He has tested them by drilling through basaltic rock, the planetary core rock, and finding sufficient petroleum to leave petroleum engineers silent in disbelief.

Much of science resolves from controversy. Heliocentricity was once a cardinal sin, as Galileo can attest. Plate tectonics was, until the early 1960s, regarded as utter nonsense. In the 1970s the rage was human activity bringing on a new Ice Age. Now, the reverse is blamed on anthropomorphics. Science is at its best when it is in turmoil. Investigate and learn more about Dr. Gold's heretical views, but do not dismiss them lightly.

Endnotes

1. PA; Act 13, Dept. pf Environmental Protection, 2012.
2. Phil Auerswald, The Coming Prosperity, Oxford University Press, 2012.
3. Although you will find a different evolutionary tale in the Appendix on Thomas Gold.
4. Op. cit. The Economist.
5. WSJ, 6/27/12, page 1.
6. BusinessDictionary.com, 2012
7. http://www.investopedia.com/terms/r/recoverabel-reserve.asp#ixzz1z9LesfN9
8. http://www.eia.gov/todayinenergy/detail.cfm?id=6990
9. http://www1.eere.energy.gov/vehiclesandfuels/facts/2011_fotw664.html
10. Modern Shale Development in the US; U.S. Dept. Energy, 4/2010: executive summary
11. http://online.wsj.com/article/SB10001424052748704740604576301550341227910.html
12. http://www.eia.gov/oiaf/aeo/otheranalysis/aeo_2010analysispapers/natgas_fuel.html
13. http://www.ogfj.com/articles/2012/06/natural.html
14. The Economist, 5/25/12. http://www.economist.com/blogs/schumpeter/2012/05/americas-falling-carbon-dioxide-emissions
15. Poland can claim 1853 as the first European oil strike and Canada hit oil in 1854; the Chinese drilled an oil well in 347 AD. And the French were using oil sands in 1498. http://www.britannica.com/blogs/2009/08/the-first-oil-well/

16. http://frac.mixplex.com/content/hydraulic-fracturing-history-enduring-technology

17. Bureau of Labor Statistics, NAICS 211-213, 10/2011

18. http://potentialgas.org/advance-summary

19. http://www.federalreserve.gov/releases/G19/Current/#releasetop

20. Thanks to the Texas State Historical Association

21. Dan Steward, The Barnett Shale Play, (ft. Worth: Ft. Worth Geological Society, 2007.

22. http://www.spe.org/jpt/print/archives/2010/12/10Hydraulic.pdf

23. http://www.visiongain.com/Report/605/The-Shale-Gas-Market-2011-2021

24. "Total reserves, production climb on mixed results," Oil and Gas Journal (December 6, 2010), pp. 46-49.

25. http://www.ogfj.com/articles/2012/03/shell-ceo-ad-dresses-ceraweek-on-the-natural-gas-revolution.html

26. http://papers.ssrn.com/sol3/papers.cfm?abstract_id=2085027

27. http://blumtexas.blogspot.com/

28. https://www.rrc.state.tx.us/barnettshale/

29. http://barnettprogress.com/powering-prosperity

30. http://www.rrc.state.tx.us/about/faqs/saltwaterwells.php

31. http://www.fountainquail.com/rover/rover.html

32. http://www.neohydro.com/index-4.html

33. http://www.halliburton.com/ps/default.aspx?pageid=4975&navid=2427

34. http://shaleoilplays.com/2011/02/number-of-new-per-mits-in-eagle-ford-shale/

35. http://ccbr.iedtexas.org/

36. http://pubs.usgs.gov/dds/dds-069/dds-069-d/

REPORTS/69_D_CH_23.pdf

37. Morehouse, David F. (July 1997). "The Intricate Puzzle of Oil & Gas "Reserve Growth"". *Natural Gas Monthly* (Energy Information Administration).

38. Engelder, Terry and Lash, Gary (2008). Unconventional Natural Gas Reservoir Could Boost U.S. Supply. Penn State Live.

39. Continental Report to Enercom, 2012, Denver, CO, 8/12, p 11.

40. http://www.tulsaworld.com/business/article.as px?subjectid=49&articleid=20110128_49_E3_ CUTLIN848378&rss_lnk=5

41. http://www.energyandcapital.com/articles/the-bakken-rail-industry/2338

42. http://www.minotdailynews.com/page/content.detail/ id/559554/The-Bakken-boom--Rail-terminal-construc-tion-picking-up-speed.html?nav=5010

43. http://www.reuters.com/article/2011/11/21/us-bakken-diesel-idUSTRE7AK0EC20111121

44. http://www.oneokpartners.com/~/media/ONEOKPart-ners/NewsRoom/PressKits/Bakken/Bakken%20Pipe-line%20FAQ.ashx

45. http://www.niobraranews.net/niobrara-decline-curve/

46. Thanks to Susan Klann at O/G Investor: http://www. tudorpickering.com/Websites/tudorpickering/Images/ News%202011/05.24.2011.Rockies.Niobrara--OGI.pdf

47. http://www.eaglelandservices.com/plays.html

48. http://www.loga.la/pdf/Economic%20Impact%20of%20 HS.pdf

49. http://www.energyindepth.org/tag/haynesville/

50. http://www.smartbrief.com/news/api/storyDetails.

jsp?issueid=E05D7254-12A8-4FDE-B8CF-
75247D50E4BB&copyid=7F97CD8D-8607-
4A2E-8710-3DCAC80EC5BC&brief=api&sb_
code=rss&&campaign=rss

51. http://www.eia.gov/todayinenergy/detail.cfm?id=570

52. http://www.energyadvancesarkansas.com/about/fayette-
ville-shale

53. http://www.energyadvancesarkansas.com/about/fayette-
ville-shale

54. http://www.eaglelandservices.com/plays.html

55. http://www.swn.com/operations/pages/fayettevilleshale.
aspx

56. http://www.solarplan.org/Research/Well%20Produc-
tion%20Profile%20for%20the%20Fayetteville%20
Shale%20Gas%20Play_Mason_OGJ_4%20April%20
2011.pdf

57. http://cber.uark.edu/Revisiting_the_Economic_Impact_
of_the_Fayetteville_Shale.pdf

58. http://www.solarplan.org/Research/Well%20Produc-
tion%20Profile%20for%20the%20Fayetteville%20
Shale%20Gas%20Play_Mason_OGJ_4%20April%20
2011.pdf

59. P 818: http://lawreview.law.uark.edu/wp-content/up-
loads/2011/02/baileyforweb.pdf

60. http://info.drillinginfo.com/urb/fayetteville/

61. http://cber.uark.edu/Revisiting_the_Economic_Impact_
of_the_Fayetteville_Shale.pdf

62. http://www.encyclopediaofarkansas.net/encyclopedia/
entry-detail.aspx?entryID=6011

63.

64. http://www.eaglelandservices.com/plays.html

65. http://www.questerre.com/en/operations/quebec/

66. http://www.ugcenter.com/Woodford/Map-Arkoma-Basin_102588

67. http://www.freerepublic.com/focus/f-news/2486859/posts

68. http://upstreampumping.com/article/well-completion-stimulation/stimulating-oklahoma-woodford-shale-fiscally-challenging-environ

69. http://www.spe.org/jpt/print/archives/2010/12/10Hydraulic.pdf

70. http://www.usoilandgas.net/acid_fracturing.htm

71. http://www.spe.org/jpt/print/archives/2010/12/10Hydraulic.pdf

72. http://www.momentivefracline.com/summer2009/index.php

73. http://www.spe.org/jpt/print/archives/2011/04/11PropantShortage.pdf

74. Recent Advances In Hydraulic Fracturing, Edited by: John L. Gidley, Stephen A. Holditch, Dale E. Nierode & Ralph W. Veatch Jr.; 1990; Society of Petroleum Engiineers (SPE)

75.

76. http://halliburtonblog.com/plug-n-perf-or-frac-values-which-is-the-best-completion-strategy/

77. http://theenergycollective.com/mark-green/68729/posted-fracturing-fluid-breakdown-energy-tomorrow-blog

78. http://www.hydraulicfracturing.com/Fracturing-Ingredients/Pages/information.aspx

79. Refer to Oil 101, Morgan Downey, Wooden Table Press, 2009 and op sit Oil and Gas for a detailed discussion of oil processing.

80. Oil and Gas Pipelines in Non-Technical Language, Meis-

ner & Leffler, Pennwell, 2006, p 170.

81. http://www.oilandgasinvestor.com/Capital-Markets-Industry-News/Effectively-Delivering-Energy-Capital-Projects-Accenture_104085

82. Op Sit,Oil and gas Pipelines, p322.

83. http://www.afdc.energy.gov/vehicles/propane.html

84. http://www.propanesafety.com/

85. A brief list of organizations trying to work in a conciliatory manner would include many others, from many walks of life, interests and industries.

86. http://www.bipartisanpolicycenter.org

87. http://www.naptp.org/documentlinks/Investor_Relations/MLP_101.pdf

88. NAPTP.org

89. Typically less than about $200,000, although you should review your tax situation with your CPA before considering placing as much as this into these tax deferred accounts. Owning a significant of the investment in any one state may lead to filing of state income taxes for earning derived from this situs.

90. The Economist, 7/28/12, p 45.

91. http://www.oilspillcommission.gov/sites/default/files/documents/OSC_Deep_Water_Summary_Recommendations_FINAL.pdf

92. B J Services, SPE.org # 119478, DVS Gupta.

93. http://www.springerlink.com/content/x001g12t2332462p/

94. Robert Skaags, President American Gas Association, address to the 1/27/10 meeting of the New York Society of Security Analysts

95. http://content.sierraclub.org/naturalgas/

96. http://stats.org/stories/2011/medias_gas_problem_oct6_11.html

97. http://www.sustainablefuture.cornell.edu/news/attachments/Howarth-EtAl-2011.pdf

98. http://iopscience.iop.org/1748-9326/6/3/034014/full-text/

99. http://www.springerlink.com/content/x001g12t2332462p/

100. http://www.ihs.com/images/MisMeasuringMethane082311.pdf

101. http://www.eia.gov/totalenergy/data/monthly/pdf/sec12_3.pdf

102. http://www.sciencenewsforkids.org/2012/07/fracing-fuels-energy-debate-and-controversy/

103. Society of Petroleum Engineers, Society of Exploration Geophysicists workshop, Broomfield, CO, 9/14/12; http://www.spe.org/events/12aden/pages/about/index.php

104. http://www.damascuscitizensforsustainability.org/2012/06/fracing-is-hardly-leakproof/. No reference is given for either of these resources from this site.

105. http://www.damascuscitizensforsustainability.org/2012/06/fracing-is-hardly-leakproof/. No reference is given for either of these resources from this site.

106. http://www.halliburton.com/public/projects/pubsdata/Hydraulic_Fracturing/fluids_disclosure.html andhttp://fracfocus.org/

107. 30 – 35% is agricultural, 15 – 30% is urban use, 15 – 30% is industrial/manufacturing, with the remainder a variety of uses. Each region has vastly different water usage tables.

108. http://www.fountainquail.com/rover/rover.html

109. http://www.neohydro.com/index-4.html

110. http://www.halliburton.com/ps/default.aspx?pageid=4975&navid=2427

111.

112. Minneapolis Star Tribune; Lee Schafer, 9/8/12

113. House subcommittee on gas prices, May 25, 2011

114. P822: http://lawreview.law.uark.edu/wp-content/uploads/2011/02/baileyforweb.pdf

115. http://www.halliburton.com/public/projects/pubsdata/Hydraulic_Fracturing/fluids_disclosure.html

116. http://www.ogfj.com/articles/2012/03/shell-ceo-addresses-ceraweek-on-the-natural-gas-revolution.html

117. http://www.worldoil.com/July-2011-Treatment-options-for-reuse-of-frac-flowback-and-produced-water-from-shale.html

118. http://www.springer.com/about+springer/media/springer+select?SGWID=0-11001-2-1128722-0

119. http://www.bls.gov/iag/tgs/iag211.htm#fatalities_injuries_and_illnesses

120. Kevin Begos, Associated Press, 7/23/12; Pittsburgh, PA

121. http://www.epmag.com/Production-Completion/DUG-Canada-2012-Demystifying-Fracing-The-Industry_102645

122. WSJ, 7/17/12, p B-1: "Chesapeake Irkes Landowners As It Renegotiates Leases".

123. http://www.nytimes.com/2011/09/03/nyregion/skepticism-directed-at-study-of-impact-of-hydraulic-fracturing.html?pagewanted=all

124. http://www.fundinguniverse.com/company-histories/ecology-and-environment-inc-history/

125. http://www.prnewswire.com/news-releases/taking-back-

americas-energy-future-from-the-lobbyists-1-million-americans-to-fight-industry-efforts-to-kill-clean-energy-agenda-160550355.html

126. The End of Country, Seamus McGraw, p. 164.

127. Comparisons here are less about size and strength than about self-imagined importance. Solyndra is no Standard Oil, by anyone's imagination. Each clearly viewed itself as the 'rising, permanent star in the firmament, only to crash Icarus-like to Earth in self –important disillusionment.

128. WSJ, 2/25/12, p A13; Holman Jenkins.

129. James Mulva, CEO Conoco-Phillips, 10/26/11, WSJ, p A13

130. http://www.eia.gov/analysis/requests/subsidy/pdf/subsidy.pdf

131. Capital, regulatory, tax, landowner and equipment costs come to mind.

132. Arthur Berman and Allen Brooks are the protagonists in this battle. Berman is the odd man out in the oil and gas industry, while Brooks is the knowledgeable insider whose comments are sought. http://www.clearonmoney.com/dw/doku.php?id=public:shale_gas_economics http://www.scribd.com/doc/92929699/Allen-Brooks-Musings-from-the-Oil-Patch-2012-05-08 http://www.pphb.com/pdfs/musings/Musings%20042611.pdf

133. Berman, A. E., 2012, U.S. Shale Gas: A Different Perspective on Future Supply and Price: Bulletin of the South Texas Geological Society (February, 2012) v. 52, no. 6, p. 19-44.

134. http://www.ogfj.com/articles/2012/06/natural.html

135. http://finance.yahoo.com/news/clean-energy-adds-major-trucking-100000728.html

136. http://www.forbes.com/sites/christopherhel-
     man/2012/06/13/shell-investing-300m-to-fuel-lng-pow-
     ered-trucks/
137. http://www.eia.gov/oiaf/aeo/otheranalysis/
     aeo_2010analysispapers/natgas_fuel.html
138. http://www.afdc.energy.gov/pdfs/final_results_wm_
     truck.pdf
139. From the author's personal visit, including the slaughter
     and feasting on a 'fat-tailed sheep'.
140. http://www.ngvamerica.org/pdfs/marketplace/
     MP.Analyses.NGVs-a.pdf
141. http://www.cngnow.com/vehicles/calculator/Pages/in-
     formation.aspx
142. http://www.cngnow.com/Pages/information.aspx
143. Tom Fowler, WSJ, 7/18/12, p R5-6.
144. http://www.newgeography.com/content/002280-the-
     explosion-oil-and-gas-extraction-jobs
145. http://thedailyreview.com/news/step-on-the-
     gas-1.1326297#
146. http://www.occeweb.com/og/oghome.htm
147. http://www.usatoday.com/money/industries/energy/
     story/2012-07-11/natural-gas-finds-lower-energy-
     costs/56157080/1
148. http://www.damascuscitizensforsustainability.
     org/2011/01/climate-benefits-of-natural-gas-may-be-
     overstated/

Index

Note: Page numbers in *italics* indicate figures or tables.

John Graves, CLU, ChFC
Editor: The Retirement Journal
Author: The 7% Solution
805.652.6948 fax: 805.652.6930

Author Biography

John Graves, ChFC, CLU has spent 26 years advising people how to become better stewards of their resources. As an independent financial advisor, he focuses on designing and maintaining clients' portfolios consistent with their needs, rather than some market paradigm. John is a Chartered Life Underwriter and Chartered Financial Consultant through The American College in Bryn Mawr, Pennsylvania.

He has traveled extensively, with more than 80 countries' stamps in his passport. His avocation is adventure. He has sailed to Hawaii several times as well as across the Atlantic and throughout the Mediterranean and Caribbean. He has trekked the Andes, the Sahara, the Taklamakan, the Serengeti, and the Namib.

In his previous career, John was a chef. He does enjoy a fine meal with a nice Bordeaux or Montalcino.

John agrees with Benjamin Graham that the search for value is far more interesting than a brief joy ride in the markets. His passion is sharing his knowledge with others so that they, too, might embrace all that life has to offer. For more information, email John at jgraves@west.net or visit www.frackusa.com.